TEACHER'S

AN INQUIRY EARTH SCIENCE PROGRAM

INVESTIGATING EARTH IN SPACE: ASTRONOMY

Ann Benbow
Mark Carpenter
Matt Hoover
Colin Mably

Developed by the American Geological Institute

Published by
It's About Time, Herff Jones Education Division, Armonk, NY

84 Business Park Drive, Armonk, NY 10504
Phone (914) 273-2233 Fax (914) 273-2227
Toll Free (888) 698-TIME
www.Its-About-Time.com

President
Tom Laster

Director of Product Develoment	**Creative Artwork**
Barbara Zahm, Ph.D.	Dennis Falcon
Project Editor	**Safety Reviewer**
Ruta Demery	Edward Robeck, Ph.D.
Creative Director/Design	**Desktop Specialist**
John Nordland	Bernardo Saravia
Associate Editor	**Technical Art**
Jane Gardner	Armstrong/Burmar

Photo Research
Bernardo Saravia, Jennifer Von Holstein

All student activities in this textbook have been designed to be as safe as possible, and have been reviewed by professionals specifically for that purpose. As well, appropriate warnings concerning potential safety hazards are included where applicable to particular activities. However, responsibility for safety remains with the student, the classroom teacher, the school principals, and the school board.

Investigating Earth Systems™ is a registered trademark of the American Geological Institute. Registered names and trademarks, etc., used in this publication, even without specific indication thereof, are not to be considered unprotected by law.

It's About Time® is a registered trademark of It's About Time, Inc. Registered names and trademarks, etc., used in this publication, even without specific indication thereof, are not to be considered unprotected by law.

© Copyright 2006: American Geological Institute

All rights reserved. No part of this publication may be reproduced, stored in a retrieval system, or transmitted, in any form or by any means, electronic, mechanical, photocopying, recording, or otherwise, without the prior written permission of the copyright owner.

Care has been taken to trace the ownership of copyright material contained in this publication. The publisher will gladly receive any information that will rectify any reference or credit line in subsequent editions.

Printed and bound in the United States of America

ISBN-10: 1-58591-266-2

ISBN-13: 978-1-58591-266-7

2 3 4 5 BR 10 09 08 07

Opinions expressed are those of the authors and not necessarily those of the donors of the American Geological Institute Foundation.

Illustrations and Photos

Student's Edition Illustrations and Photos

A60 (Middle and Bottom Left), A63 (Bottom Left Composite and Middle Right) Anglo-Australian Observatory/David Malin

A6 (Bottom), A12, A22, A23, A24 (Bottom), A39, A40 (Top and Bottom), A41 (Bottom), A42 (Top), A46, A53 (Top and Middle), A65 by Stuart Armstrong

Av (Top and Bottom), Axii, A2, A10, A16, A17, A28, A36, A37, A47, A48, illustrations by Dennis Falcon

A68 (Top) Fermilab Visual Media Services

A38 Getty Images

Axi (Bottom Right), A68 (Bottom) Griffith Observatory/Anthony Cook

A24 (Top) Tom McGuire

A70 (Top) Courtesy of the NAIC - Arecibo Observatory, a facility of the NSF

Axi (Top Right and Bottom Left), A4, A5, A6 (Top), A9, A15, A30 (Top and Bottom), A33, A41 (Top), A42 (Left), A43 (Top and Bottom), A53 (Bottom), A69, A71 (Top and Bottom), NASA

A66 NASA, Wolfgang Brandner JPL-IPAC, Eva K. Grebel

A60 (Top Right), A63 (Top and Bottom Left Composite) NASA/CXX/M. Weiss

A1 NASA Goddard Space Flight Center

A61 (Bottom Left), 67 NASA, The Hubble Heritage Team (STScl)

A26, A51 (Bottom Left and Right), A70 (Bottom) NASA Jet Propulsion Laboratory (NASA-JPL)

A42 (Top) NASA, Courtesy McRel

A55 NASA, Robert Williams and the Hubble Deep Field Team (STScl)

A64 (Left and Middle) NASA/WMAP Science Team

A8 Courtesy NASCO

Axi (Top Left), A8 (Top), A13, A14, A44, A50, A51 (Top), A52, A57, A60 (Middle Right), A61 (Top, Middle and Bottom Right), A62, A73, A74 (Top and Middle) Photodisc

A29 Wojtek Rychlik

Investigating Earth Systems Team

Curriculum Developers

Ann Benbow, Ph.D., Director of Education, Outreach and Development
    American Geological Institute
Mark Carpenter, Education Specialist
    American Geological Institute
Matthew Hoover, Education Specialist
    American Geological Institute
Colin Mably, Senior Advisor – Communications Technology
    American Geological Institute

Content Reviewers

David Gordon, Ph.D.
Eric L. Winter, Ph.D.

Original Project Personnel

Michael J. Smith, Principal Investigator
Colin Mably, Project Director
John B. Southard, Senior Writer
William O. Jones, Contributing Writer
Caitlin N. Callahan, Project Assistant
William S. Houston, Field Test Coordinator
Harvey Rosenbaum, Field Test Evaluator
Fred Finley, Project Evaluator
Lynn Lindow, Pilot Test Evaluator
Robert L. Heller, Principal Investigator
Charles Groat, Principal Investigator
Robert Ridky, Project Director
Marilyn Suiter, Co-Principal Investigator

Advisory Board

National Advisory Board

Jane Crowder
    Middle School Teacher, WA
Kerry Davidson
    Louisiana Board of Regents, LA
Joseph D. Exline
    Educational Consultant, VA
Louis A. Fernandez
    California State University, CA
Frank Watt Ireton
    National Earth Science Teachers Association, DC
LeRoy Lee
    Wisconsin Academy of Sciences, Arts and Letters, WI
Donald W. Lewis
    Chevron Corporation, CA
James V. O'Connor (deceased)
    University of the District of Columbia, DC
Roger A. Pielke Sr.
    Colorado State University, CO
Dorothy Stout (deceased)
    Cypress College, CA
Lois Veath
    Advisory Board Chairperson - Chadron State College, NE

National Science Foundation Program Officers

Gerhard Salinger
Patricia Morse

The Development Team for Investigating Earth in Space: Astronomy

Ann Benbow, Ph.D. is a researcher, curriculum developer, teacher, and trainer. She is currently Director of Education, Development and Outreach at AGI. After teaching science (biology, chemistry, and Earth science) in high school, elementary school, and two-year college, she taught elementary and secondary science methods at the university level. She worked in research and development for the Education Division of the American Chemical Society for over 12 years. During that time, she directed a number of national educational grants from the National Science Foundation (NSF). Her work in the informal science arena included a period of time as managing editor of *WonderScience* magazine for children and adults, and administrator for the Parents and Children for Terrific Science mini-grant program. Dr. Benbow is currently Principal Investigator of two NSF-supported projects, has co-authored a college textbook on elementary science methods for Wadsworth Publishing, and recently published a book on improving communication techniques with adult learners. Dr. Benbow has a B.S. in Biology from St. Mary's College in Maryland, an M.Ed. in Science Education, and a Ph.D. in Curriculum and Instruction from the University of Maryland College Park.

Mark Carpenter is an Education Specialist at the American Geological Institute. After receiving a B.S. in geology from Exeter University, England he undertook a graduate degree at the University of Waterloo and Wilfrid Laurier, Canada, where he began designing geology investigations for undergraduate students and worked as an instructor. He has worked in basin hydrology in Ontario, Canada, and studied mountain geology in the Pakistan, and the Nepal Himalayas. As a designer of learning materials for AGI, he has made educational films to support teachers and is actively engaged in designing inquiry-based activities in Earth system science for middle school children in the United States.

Matthew Hoover serves as Education Specialist for the American Geological Institute, developing Earth science educational resources and curriculum programs at the middle and high school levels. He received a B.S. degree in Geology from Boston College, an M.A. degree in Environmental Policy from George Washington University and an M.Ed. in Curriculum and Instruction from George Mason University. As a certified teacher, he has taught elementary and middle school Earth, life, and physical sciences. Prior to joining AGI, he worked for NASA's GLOBE Program, coordinating teacher trainings and designing environmental science investigations and learning activities for K-12 students.

Colin Mably is a curriculum developer, designer/illustrator, educational television producer, teacher, and trainer. He currently acts as Senior Advisor for Communications to AGI. After ten years as an elementary and middle school teacher, he joined the faculty of Furzedown College of Education and later became Principal Lecturer in the School of Education at the University of East London. Leaving academia to form an educational multimedia company, he developed video-based elementary science and mathematics curricula, in the UK and the USA. He has been a key curriculum developer for several NSF-funded national curriculum projects at middle, high school, and college levels. For AGI, he directed the design and development of the IES curriculum and also training workshops for pilot and field-test teachers. He has also recently co-authored a college textbook on elementary science methods. He received certified teacher status from Oxford University Institute of Education, and an Advanced Diploma in Education from London University Institute of Education.

Acknowledgements

Project Team
Marcus Milling
Executive Director - AGI, VA
Michael Smith
Principal Investigator - Director of Education - AGI, VA
Colin Mably
Project Director/Curriculum Designer - Educational Visions, MD
Matthew Smith
Project Coordinator
Program Manager - AGI, VA
Fred Finley
Project Evaluator
University of Minnesota, MN
Joe Moran
American Meteorological Society
Lynn Lindow
Pilot Test Evaluator
University of Minnesota, MN
Harvey Rosenbaum
Field Test Evaluator
Montgomery School District, MD
Ann Benbow
Project Advisor - American Chemical Society, DC
Robert Ridky
Original Project Director
University of Maryland, MD
Chip Groat
Original Principal Investigator - University of Texas
El Paso, TX
Marilyn Suiter
Original Co-principal Investigator - AGI, VA
William Houston
Field Test Manager
Caitlin Callahan - Project Assistant

Original and Contributing Authors
Oceans
George Dawson
Florida State University, FL
Joseph F. Donoghue
Florida State University, FL
Ann Benbow
American Chemical Society
Michael Smith
American Geological Institute
Soil
Robert Ridky
University of Maryland, MD
Colin Mably - LaPlata, MD
John Southard
Massachusetts Institute of Technology, MA
Michael Smith
American Geological Institute
Fossils
Robert Gastaldo
Colby College, ME
Colin Mably - LaPlata, MD
Michael Smith
American Geological Institute
Climate and Weather
Mike Mogil
How the Weather Works, MD
Ann Benbow
American Chemical Society
Joe Moran
American Meteorological Society
Michael Smith
American Geological Institute
Energy Resources
Laurie Martin-Vermilyea
American Geological Institute
Michael Smith
American Geological Institute
Our Dynamic Planet
Michael Smith
American Geological Institute
Rocks and Landforms
Michael Smith
American Geological Institute
Water as a Resource
Ann Benbow
American Chemical Society
Michael Smith
American Geological Institute
Materials and Minerals
Mary Poulton
University of Arizona, AZ
Colin Mably - LaPlata, MD
Michael Smith
American Geological Institute
Earth in Space: Astronomy
Ann Benbow
American Geological Institute
Mark Carpenter
American Geological Institute
Matthew Hoover
American Geological Institute
Colin Mably
American Geological Institute

Advisory Board
Jane Crowder
Middle School Teacher, WA
Kerry Davidson
Louisiana Board of Regents, LA
Joseph D. Exline
Educational Consultant, VA
Louis A. Fernandez
California State University, CA
Frank Watt Ireton
National Earth Science Teachers Association, DC
LeRoy Lee
Wisconsin Academy of Sciences, Arts and Letters, WI
Donald W. Lewis
Chevron Corporation, CA
James V. O'Connor (deceased)
University of the District of Columbia, DC
Roger A. Pielke Sr.
Colorado State University, CO
Dorothy Stout (deceased)
Cypress College, CA
Lois Veath
Advisory Board Chairperson
Chadron State College, NE

Pilot Test Teachers
Debbie Bambino
Philadelphia, PA
Barbara Barden - Rittman, OH
Louisa Bliss - Bethlehem, NH
Mike Bradshaw - Houston, TX
Greta Branch - Reno, NV
Garnetta Chain - Piscataway, NJ
Roy Chambers Portland, OR
Laurie Corbett - Sayre, PA
James Cole - New York, NY
Collette Craig - Reno, NV
Anne Douglas - Houston, TX
Jacqueline Dubin - Roslyn, PA
Jane Evans - Media, PA
Gail Gant - Houston, TX
Joan Gentry - Houston, TX
Pat Gram - Aurora, OH
Robert Haffner - Akron, OH
Joe Hampel - Swarthmore, PA
Wayne Hayes - West Green, GA
Mark Johnson - Reno, NV
Cheryl Joloza - Philadelphia, PA
Jeff Luckey - Houston, TX
Karen Luniewski
Reistertown, MD
Cassie Major - Plainfield, VT
Carol Miller - Houston, TX
Melissa Murray - Reno, NV
Mary-Lou Northrop
North Kingstown, RI
Keith Olive - Ellensburg, WA
Tracey Oliver - Philadelphia, PA
Nicole Pfister - Londonderry, VT
Beth Price - Reno, NV
Joyce Ramig - Houston, TX
Julie Revilla - Woodbridge, VA
Steve Roberts - Meredith, NH
Cheryl Skipworth
Philadelphia, PA
Brent Stenson - Valdosta, GA
Elva Stout - Evans, GA
Regina Toscani
Philadelphia, PA
Bill Waterhouse
North Woodstock, NH
Leonard White
Philadelphia, PA
Paul Williams - Lowerford, VT
Bob Zafran - San Jose, CA
Missi Zender - Twinsburg, OH

Field Test Teachers
Eric Anderson - Carson City, NV
Katie Bauer - Rockport, ME
Kathleen Berdel - Philadelphia, PA
Wanda Blake - Macon, GA
Beverly Bowers
Mannington, WV
Rick Chiera - Monroe Falls, OH
Don Cole - Akron, OH
Patte Cotner - Bossier City, LA

Johnny DeFreese - Haughton, LA
Mary Devine - Astoria, NY
Cheryl Dodes - Queens, NY
Brenda Engstrom - Warwick, RI
Lisa Gioe-Cordi - Brooklyn, NY
Pat Gram - Aurora, OH
Mark Johnson - Reno, NV
Chicory Koren - Kent, OH
Marilyn Krupnick
Philadelphia, PA
Melissa Loftin - Bossier City, LA
Janet Lundy - Reno, NV
Vaughn Martin - Easton, ME
Anita Mathis - Fort Valley, GA
Laurie Newton - Truckee, NV
Debbie O'Gorman - Reno, NV
Joe Parlier - Barnesville, GA
Sunny Posey - Bossier City, LA
Beth Price - Reno, NV
Stan Robinson
Mannington, WV
Mandy Thorne
Mannington, WV
Marti Tomko
Westminster, MD
Jim Trogden - Rittman, OH

Torri Weed - Stonington, ME
Gene Winegart - Shreveport, LA
Dawn Wise - Peru, ME
Paula Wright - Gray, GA

Field Test Teachers and Specialists

Jenny Soro; Ruby Everage
Daniel Boone Elementary
Joyce Anderson; Maureen Tucker
Burnside Scholastic Academy
Aimee Ray; Marvin Nochowitz
Haines Elementary
Nicole Hauser; Noreen Sepulveda
Healy Elementary
Roseann Pavelka; Ann Doyle
Kinzie Elementary
Katherine Lee
McCorkle Elementary
Terri Zachary; Chandra Garcia
O'Toole Elementary
Brenda Armstrong; Delores McKinney
Overton Elementary
Kathryn Doyle; Patsy Moore
Pirie Magnet Elementary
Veronica Johnson ; Constance Grimm-Grason
Ray Elementary

Mary Pat Robertson; Constance Grimm-Grason
Ray Elementary
Raymond Montes; Barbara Dubielak-Wood
Reilly Elementary
Raul Bermejo; Tammy Valaveris
Columbia Explorers Academy
Kim John-Baptiste; Lillian Degand
Finkl Elementary
Marie Clouston
Peck Elementary

Facilitators

Linda Carter, Gary Morrissey, Alan Nelson
Office of Math and Science for Chicago Public Schools

IMPORTANT NOTICE

The *Investigating Earth Systems*™ series of modules is intended for use by students under the direct supervision of a qualified teacher. The experiments described in this book involve substances that may be harmful if they are misused or if the procedures described are not followed. Read cautions carefully and follow all directions. Do not use or combine any substances or materials not specifically called for in carrying out experiments. Other substances are mentioned for educational purposes only and should not be used by students unless the instructions specifically indicate.

The materials, safety information, and procedures contained in this book are believed to be reliable. This information and these procedures should serve only as a starting point for classroom or laboratory practices, and they do not purport to specify minimal legal standards or to represent the policy of the American Geological Institute. No warranty, guarantee, or representation is made by the American Geological Institute as to the accuracy or specificity of the information contained herein, and the American Geological Institute assumes no responsibility in connection therewith. The added safety information is intended to provide basic guidelines for safe practices. It cannot be assumed that all necessary warnings and precautionary measures are contained in the printed material and that other additional information and measures may not be required.

This work is based upon work supported by the National Science Foundation under Grant No. 9353035 with additional support from the Chevron Corporation. Any opinions, findings, and conclusions or recommendations expressed in this publication are those of the authors and do not necessarily reflect the views of the National Science Foundation or the Chevron Corporation. Any mention of trade names does not imply endorsement from the National Science Foundation or the Chevron Corporation.

Table of Contents

Investigating Earth Systems Team iv
Acknowledgements vi
Developing *Investigating Earth Systems* x
Investigating Earth Systems Modules xi
Investigating Earth Systems:
    Correlation to the National Science Education Standards xii
Using *Investigating Earth Systems* Features in Your Classroom xiv
Using the *Investigating Earth Systems* Web Site xxviii
Enhancing Teacher Content Knowledge xxix
Managing Inquiry in Your *Investigating Earth Systems* Classroom xxx
Assessing Student Learning in *Investigating Earth Systems* xxxiii
Investigating Earth Systems Assessment Tools xxxvi
Reviewing and Reflecting upon Your Teaching xl
Investigating Earth in Space: Astronomy: Introduction 1
Students' Conceptions about Earth in Space: Astronomy 3
Investigating Earth in Space: Astronomy: Module Flow 6
Investigating Earth in Space: Astronomy: Module Objectives 7
National Science Education Content Standards 10
Key NSES Earth Science Standards Addressed in *IES* Earth in Space: Astronomy ... 11
Materials and Equipment List for Investigating Earth in Space: Astronomy 13
Teaching the Nature of Science 16
Pre-assessment 18
Introducing the Earth System 23
Introducing Inquiry Processes 27
Introducing Earth in Space: Astronomy 29
Why is Astronomy Important in Studying Earth and Space? 31
Investigation 1: There's No Place Like Home 33
Investigation 2: The Earth's Moon 51
Investigation 3: The Earth and Its Moon 73
Investigation 4: Finding Our Place in Space 101
Investigation 5: The Sun and Its Central Role in Our Solar System 127
Investigation 6: The Planetary Council 165
Investigation 7: What is Beyond Our Solar System? 191
Investigation 8: Discovering the Difference Between
    Science Fact and Science Fiction 235
Reflecting 250
Appendices: Alternative End-of-Module Assessment 254
                Assessment Tools 258
                Blackline Masters 268

Developing Investigating Earth Systems

Welcome to *Investigating Earth Systems (IES)! IES* was developed through funding from the National Science Foundation and the American Geological Institute Foundation. Classroom teachers, scientists, and thousands of students across America helped to create *IES*. In the 1997-98 school year, scientists and curriculum developers drafted nine *IES* modules. They were pilot tested by 43 teachers in 14 states from Washington to Georgia. Faculty from the University of Minnesota conducted an independent evaluation of the pilot test in 1998, which was used to revise the program for a nationwide field test during the 1999-2000 school year. A comprehensive evaluation of student learning by a professional field-test evaluator showed that *IES* modules led to significant gains in student understanding of fundamental Earth science concepts. Field-test feedback from 34 teachers and content reviews from 33 professional Earth scientists were used to produce the commercial edition you have selected for your classroom.

Inquiry and the interrelation of Earth's systems form the backbone of *IES*. Often taught as a linear sequence of events called "the scientific method," inquiry underlies all scientific processes and can take many different forms. It is very important that students develop an understanding of inquiry processes as they use them. Your students naturally use inquiry processes when they solve problems. Like scientists, students usually form a question to investigate after first looking at what is observable or known. They predict the most likely answer to a question. They base this prediction on what they already know to be true. Unlike professional scientists, your students may not devote much thought to these processes. In order to be objective, students must formally recognize these processes as they do them. To make sure that the way they test ideas is fair, scientists think very carefully about the design of their investigations. This is a skill your students will practice throughout each *IES* module.

All *Investigating Earth Systems* modules also encourage students to think about the Earth as a system. Upon completing each investigation they are asked to relate what they have learned to the Earth Systems (see the *Earth System Connection* sheet in the **Appendix**). Integrating the processes of the biosphere, geosphere, hydrosphere, and atmosphere will open up a new way of looking at the world for most students. Understanding that the Earth is dynamic and that it affects living things, often in unexpected ways, will engage them and make the topics more relevant.

We trust that you will find the Teacher's Edition that accompanies each student module to be useful. It provides **Background Information** on the concepts explored in the module, as well as strategies for incorporating inquiry and a systems-based approach into your classroom. Enjoy your investigation!

Investigating Earth Systems Modules

Climate and Weather

Energy Resources

Fossils

Materials and Minerals

Oceans

Our Dynamic Planet

Rocks and Landforms

Soil

Water as a Resource

Earth in Space: Astronomy

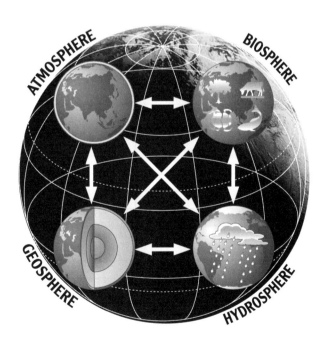

Investigating Earth Systems: Correlation to the National Science Education Standards

National Science Education Content Standards Grades 5 – 8

	Soil	Rocks and Landforms	Oceans	Climate and Weather	Our Dynamic Planet	Materials and Minerals	Energy Resources	Water as a Resource	Fossils	Astronomy
UNIFYING CONCEPTS AND PROCESSES										
System, order and organization	•	•	•	•	•	•	•	•	•	•
Evidence, models, and explanation	•	•	•	•	•	•	•	•	•	•
Constancy, change, and measurement	•	•	•	•	•	•	•	•	•	•
Evolution and equilibrium		•	•	•	•			•	•	•
Form and function									•	
SCIENCE AS INQUIRY										
Identify questions that can be answered through scientific investigations	•	•	•	•	•	•	•	•	•	•
Design and conduct scientific investigations	•	•	•	•	•	•	•	•	•	•
Use tools and techniques to gather, analyze, and interpret data	•	•	•	•	•	•	•	•	•	•
Develop descriptions, explanations, predictions and models based on evidence	•	•	•	•	•	•	•	•	•	•
Think critically and logically to make the relationships between evidence and explanation	•	•	•	•	•	•	•	•	•	•
Recognize and analyze alternative explanations and predictions	•	•	•	•	•	•	•	•	•	•
Communicate scientific procedures and explanations	•	•	•	•		•	•	•	•	•
Use mathematics in all aspects of scientific inquiry	•	•	•	•	•	•	•	•	•	•
Understand scientific inquiry	•	•	•	•	•	•	•	•	•	•
PHYSICAL SCIENCE										
Properties and Changes of Properties in Matter	•	•	•		•	•	•	•		•
Motions and Forces	•		•							•
Transfer of Energy		•	•	•	•	•	•	•		•
LIFE SCIENCE										
Populations and Ecosystems			•				•	•	•	
Diversity and Adaptation of Organisms			•		•				•	•

© It's About Time

Investigating Earth Systems: Correlation to the National Science Education Standards

National Science Education Content Standards Grades 5 – 8	Soil	Rocks and Landforms	Oceans	Climate and Weather	Our Dynamic Planet	Materials and Minerals	Energy Resources	Water as a Resource	Fossils	Astronomy
EARTH AND SPACE SCIENCE										
Structure of the Earth system	•	•	•	•	•	•	•	•	•	•
Earth's History	•	•	•	•	•	•	•	•	•	•
Earth in the Solar System			•	•	•		•	•		•
SCIENCE AND TECHNOLOGY										
Abilities of technological design	•	•	•	•	•	•		•		•
Understandings about science and technology		•	•			•	•	•		•
SCIENCE IN PERSONAL AND SOCIAL PERSPECTIVES										
Personal health	•							•		
Populations, resources, and environment	•					•	•	•		
Natural Hazards		•		•	•	•				•
Risks and benefits				•			•			•
Science and technology in society	•	•	•	•	•	•	•	•	•	•
HISTORY AND NATURE OF SCIENCE										
Science as a human endeavor	•	•	•	•	•	•	•	•	•	•
Nature of science	•	•	•	•	•	•	•	•	•	•
History of science			•		•				•	•

Using Investigating Earth Systems Features in Your Classroom

1. Pre-assessment

Designed under the umbrella framework of "science for all students," meaning that all students should be able to engage in inquiry and learn core science concepts, *Investigating Earth Systems* helps you to tailor instruction to meet your students' needs. A crucial first step in this framework is to ascertain what knowledge, experience, and understanding your students bring to their study of a module. The pre-assessment consists of four questions geared to the major concepts and understandings targeted in the unit. Students write and draw what they know about the major topics and concepts. This information is recorded and shared in an informal discussion prior to engaging in hands-on inquiry. The discussion enables students to recognize how much there is to learn and appreciate, and that by exploring the unit together, the entire classroom can emerge from the experience with a better understanding of core concepts and themes. Students' responses provide crucial pre-assessment data for you. By examining their written work and probing for further detail during the classroom conversation, you can identify strengths and weaknesses in students' understandings, as well as their abilities to communicate that understanding to others. It is important that the pre-assessment not be viewed as a test, and that judgments about the accuracy of responses not be evaluated in writing or through your comments during the conversation. The goal is to ascertain and probe, not judge, and to create a safe classroom environment in which students feel comfortable sharing their ideas and knowledge. Students revisit these pre-assessment questions informally throughout the unit. At the end of the unit, students respond to the same four questions in the section called **Back to the Beginning**. The pre-assessment thus helps you and your students to make judgments about their growth in understanding and ability throughout the module.

Investigating Earth Systems

2. The Earth System

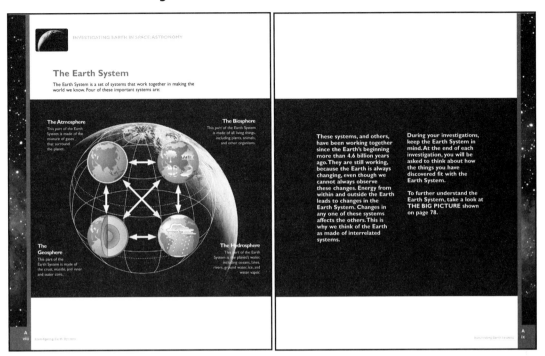

National Science Education Standards link...

"A major goal of science in the middle grades is for students to develop an understanding of Earth (and the solar system) as a set of closely coupled systems. The idea of systems provides a framework in which students can investigate the four major interacting components of the Earth System – geosphere (crust, mantle, and core), hydrosphere (water), atmosphere (air), and the biosphere (the realm of living things)."

NSES content standard D "Developing Student Understanding" (pages 158-159)

Understanding the Earth system is an overall goal of the *Investigating Earth Systems* series. It is a difficult and complex set of concepts to grasp, because it is inferred rather than observed directly. Yet even the smallest component of Earth science can be linked to the Earth system. As your students progress through each module, an increasing number of connections with the Earth system will arise. Your students may not, however, immediately see these connections. At the end of every investigation, they will be asked to link what they have discovered with ideas about the Earth system. They will also be asked to write about this in their journals. A **Blackline Master** (*Earth System Connection* sheet) is available in each Teacher's Edition. Students can use this to record connections that they make as they complete each investigation. At the very end of the module they will be asked to review everything they have learned in relation to the Earth system. The aim is for students to have a working understanding of the Earth System by the time they complete grade 8. They will need your help accomplishing this.

For example, in *Investigating Rocks and Landforms*, students work with models to simulate Earth processes, such as erosion of stream sediment and deposition of that sediment on floodplains and in deltas. Changes in inputs in one part of the system (say rainfall, from the atmosphere), affect other parts of the system (stream flows, erosion on river bends, amount of sediment carried by the stream, and deposition of sediment on floodplains or in deltas). These changes affect, in turn, other parts of the system (for example, floods that affect human populations, i.e., the biosphere). In the same module, students explore the rock record within their community and develop understandings about how interactions between the hydrosphere, atmosphere, geosphere, and biosphere change the landscape over time. These are just some of the many ways that *Investigating Earth Systems* modules foster and promote student thinking about the dynamic nature and interactions of Earth systems—biosphere, geosphere, atmosphere, and hydrosphere.

3. Introducing Inquiry Processes

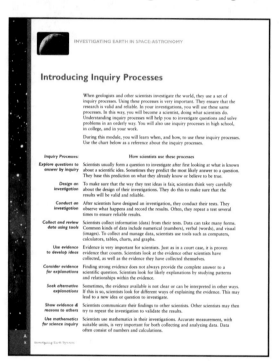

Inquiry is at the heart of *Investigating Earth Systems*. That is why each module title begins with the title "Investigating." In the National Science Education Standards, inquiry is the first Content Standard. NSES then lists a range of points about inquiry. These fundamental components of inquiry were written into the list shown at the beginning of each student module. It is very important that students be reminded of the steps in the inquiry process as they perform them. Inquiry depends on active student participation. Ideas on how to make inquiry successful in the classroom appear throughout the modules and in the "Managing Inquiry in Your *Investigating Earth Systems* Classroom" section of this Teacher's Edition.

It is very important that students develop an understanding of the inquiry processes as they use them. Stress the importance of inquiry processes as they occur in your investigations. Provoke students to think about why these processes are important. Collecting good data, using evidence, considering alternative explanations, showing evidence to others, and using mathematics are all essential to *IES*. Use examples to demonstrate these processes whenever possible. At the end of every investigation, students are asked to reflect on the scientific inquiry processes they used. Refer students to the list of inquiry processes on page x of the Student Book as they think about scientific inquiry and answer the questions.

4. Introducing the Module

Each *IES* module begins with photographs and questions. This is an introduction to the module for your students. It is designed to give them a brief overview of the content of the module and set their investigations into a relevant and meaningful context. Students will have had a variety of experiences with the content of the module. This is an opportunity for them to offer some of their own experiences in a general discussion, using these questions as prompts. This section of each *IES* module follows the pre-assessment, where students spend time thinking about what they already know about the content of the module. The photographs and questions can be used to focus the students' thinking.

The ideas students share in the introduction to the module provide you with additional pre-assessment data. The experiences they describe and the way in which they are discussed will alert you to their general level of understanding about these topics. To encourage sharing and to provide a record, teachers find it useful to quickly summarize the main points that emerge from discussion. You can do this on the chalkboard or flipchart for all to see. This can be displayed as students work through the module and added to with each new experience. For your own assessment purposes, it will be useful to keep a record of these early indicators of student understanding.

5. Key Question

Each *Investigating Earth Systems* investigation begins with a **Key Question** – an open-ended question that gives teachers the opportunity to explore what their students know about the central concepts of the activity. Uncovering students' thinking (their prior knowledge) and exposing the diversity of ideas in the classroom are the first steps in the learning cycle. One of the most fundamental principles derived from many years of research on student learning is that:

"Students come to the classroom with preconceptions about how the world works. If their initial understanding is not engaged, they may fail to grasp the new concepts and information that are taught, or they may learn them for the purposes of a test but revert to their preconceptions outside the classroom." (*How People Learn: Bridging Research and Practice*, National Research Council, 1999, P. 10.)

This principle has been illustrated through the *Private Universe* series of videotapes that show Harvard graduates responding to basic science questions in much the same way that fourth grade students do. Although the videotapes revealed that the Harvard graduates used a more sophisticated vocabulary, the majority held onto the same naïve, incorrect conceptions of elementary school students. Research on learning suggests that the belief systems of students who are not confronted with what they believe and adequately shown why they should give up that belief system remain intact. Real learning requires confronting one's beliefs and testing them in light of competing explanations.

Drawing out and working with students' preconceptions is important for learners. In *Investigating Earth Systems*, the **Key Question** is used to ascertain students' prior knowledge about the key concept or Earth science processes or events explored in the activity. Students verbalize what they think about the age of the Earth, the causes of volcanoes, or the way that the landscape changes over time before they embark on an activity designed to challenge and test these beliefs. A brief discussion about the diversity of beliefs in the classroom makes students consider how their ideas compare to others and the evidence that supports their view of volcanoes, earthquakes, or seasons.

Investigating Earth Systems

The **Key Question** is not a conclusion, but a lead into inquiry. It is not designed to instantly yield the "correct answer" or a debate about the features of the question, or to bring closure. The activity that follows will provide that discussion as students analyze and discuss the results of inquiry. Students are encouraged to record their ideas in words and/or drawings to ensure that they have considered their prior knowledge. After students discuss their ideas in pairs or in small groups, teachers activate a class discussion. A discussion with fellow students prior to class discussion may encourage students to exchange ideas without the fear of personally giving a "wrong answer." Teachers sometimes have students exchange papers and volunteer responses that they find interesting.

Some teachers prefer to have students record their responses to these questions. They then call for volunteers to offer ideas up for discussion. Other teachers prefer to start with discussion by asking students to volunteer their ideas. In either situation, it is important that teachers encourage the sharing of ideas by not judging responses as "right" or "wrong." It is also important that teachers keep a record of the variety of ideas, which can be displayed in the classroom (on a sheet of easel pad paper or on an overhead transparency) and referred to as students explore the concepts in the module. Teachers often find that they can group responses into a few categories and record the number of students who hold each idea. The photograph in each **Key Question** section was designed to stimulate student thinking and help students to make the specific kinds of connections emphasized in each activity.

6. Investigate

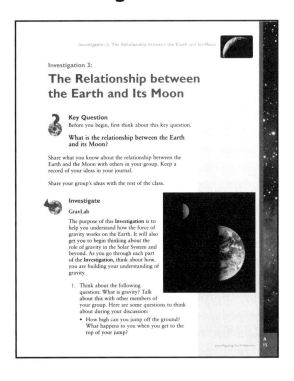

Investigating Earth Systems is a hands-on, minds-on curriculum. In designing *Investigating Earth Systems*, we were guided by the research on learning, which points out how important ***doing*** Earth Science is to ***learning*** Earth Science. Testing of *Investigating Earth Systems* activities by teachers across America provided critical testimonial and quantitative measures of the importance of the activities to student learning. In small groups and as a class, students take part in doing hands-on experiments, participating in field work, or searching for answers using the Internet and reference materials. **Blackline Masters** are included in the Teacher's Editions for any maps or illustrations that are essential for students to complete the activity.

Each part of an *Investigating Earth Systems* investigation, as well as the sequence of activities within a module, moves from concrete to abstract. Hands-on activities provide the basis for exploring student beliefs about how the world works and to manipulate variables that affect the outcomes of experiments, models, or simulations. Later in each activity, formal labels are applied to concepts by introducing terminology used to describe the processes that students have explored through hands-on activity. This flow from concrete (hands-on) to abstract (formal explanations) is progressive – students begin to develop their own explanations for phenomena by responding to questions within the **Investigate** section.

Each activity has instructions for each part of the investigation. Materials kits are available for purchase, but you will also need to obtain some resources from outside suppliers, such as topographic and geologic maps of your community, state, or region. The *Investigating Earth Systems* web site will direct you to sources where you can gather such materials.

Most **Investigate** activities will require between one and two class periods. The variety of school schedules and student needs makes it difficult to predict exactly how much time your class will need. For example, if students need to construct a graph for part of an investigation, and the students have never been exposed to graphing, then this investigation may require additional time and could become part of a mathematics lesson.

The most challenging aspect of *Investigating Earth Systems* for teachers to "master" is that the **Investigate** section of each activity has been designed to be student-driven. Students learn more when they have to struggle to "figure things out" and work in collaborative groups to solve problems as a team. Teachers will have to resist the temptation to provide the answers to students when they get "stuck" or hung up on part of a problem. Eventually, students learn that while they can call upon their teacher for assistance, the teacher is not going to "show them the answer." Field testing of *Investigating Earth Systems* revealed that teachers who were most successful in getting their students to solve problems as a team were patient with this process and steadfast in their determination to act as facilitators of learning during the **Investigate** portion of activities. As one teacher noted, "My response to questions during the investigation was like a mantra, 'What do you think you need to do to solve this?' My students eventually realized that although I was there to provide guidance, they weren't going to get the solution out of me."

Another concern that many teachers have when examining *Investigating Earth Systems* for the first time is that their students do not have the background knowledge to do the investigations. They want to deliver a lecture about the phenomena before allowing students to do the investigation. Such an approach is common to many traditional programs and is inconsistent with the pedagogical theory used to design *Investigating Earth Systems*. The appropriate place for delivering a lecture or reading text in *Investigating Earth Systems* is following the investigation, not preceding it.

For example, suppose a group of students has been asked to interpret a map. The traditional approach to science education is for the teacher to give a lecture or assign a reading, "How to Interpret Maps," then give students practice reading maps. *Investigating Earth Systems* teachers recognize that while students may lack some specific skills (reading latitude and longitude, for example), within a group of four students, it is not uncommon for at least one of the students to have a vital skill or piece of knowledge that is required to solve a problem. The one or two students who have been exposed to (or better yet, understood) latitude and longitude have the opportunity to shine within the group by contributing that vital piece of information or demonstrating a skill. That's how scientific research teams work – specialists bring expertise to the group, and by working together, the group achieves something that no one could achieve working alone. The **Investigate** section of *Investigating Earth Systems* is modeled in the spirit of the scientific research team.

7. Inquiry

Inquiry is the first content standard in the National Science Education Standards (NSES). The American Association for the Advancement of Science's (AAAS) Benchmarks for Science Literacy also places considerable emphasis on scientific inquiry (see excerpts on the following page). *IES* has been designed to remind students to reflect on inquiry processes as they carry out their investigations. The student journal is an important tool in helping students to develop these understandings. In using the journal, students are modeling what scientists do. Your students are young scientists as they investigate Earth science questions. Encourage your students to think of themselves in this way and to see their journals as records of their investigations.

Inquiry
Representing Information

Communicating findings to other scientists is very important in scientific inquiry. In this investigation it is important for you to find good ways of showing what you learned to others in your class. Be sure your maps and displays are clearly labeled and well organized.

An icon was developed to draw students' attention to brief descriptions of inquiry processes in the margins of the student module. The icon and explanations provide opportunities to direct students' attention to what they are doing, and thus serve as an important metacognitive tool to stimulate thinking about thinking.

National Science Education Standards link...

Content Standard A
As a result of activities in grades 5-8, all students should develop:
- Abilities necessary to do scientific inquiry
- Understandings about scientific inquiry

Abilities Necessary to do Scientific Inquiry
- Identify questions that can be answered through scientific investigations
- Use appropriate tools and techniques to gather, analyze, and interpret data
- Develop descriptions, explanations, predictions, and models using evidence
- Think critically and logically to make the relationships between evidence and explanations
- Recognize and analyze alternative explanations and predictions
- Communicate scientific procedures and explanations
- Use mathematics in all aspects of scientific inquiry

(From National Science Education Standards, pages 145-148)

Benchmarks for Science Literacy link...

The Nature of Science Inquiry: Grades 6 through 8
- At this level, students need to become more systematic and sophisticated in conducting their investigations, some of which may last for several weeks. That means closing in on an understanding of what constitutes a good experiment. The concept of controlling variables is straightforward, but achieving it in practice is difficult. Students can make some headway, however, by participating in enough experimental investigations (not to the exclusion, of course, of other kinds of investigations) and explicitly discussing how explanation relates to experimental design.

- Student investigations ought to constitute a significant part—but only a part—of the total science experience. Systematic learning of science concepts must also have a place in the curriculum, for it is not possible for students to discover all the concepts they need to learn, or to observe all of the phenomena they need to encounter, solely through their own laboratory investigations. And even though the main purpose of student investigations is to help students learn how science works, it is important to back up such experience with selected readings. This level is a good time to introduce stories (true and fictional) of scientists making discoveries – not just world-famous scientists, but scientists of very different backgrounds, ages, cultures, places, and times.

(From Benchmarks for Science Literacy, page 12)

Investigating Earth Systems

8. Digging Deeper

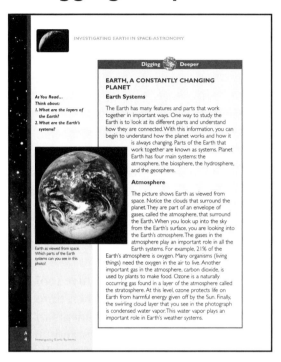

This section provides text, illustrations, data tables, and photographs that give students greater insight into the concepts explored in the activity. Teachers often assign **As You Read** questions as homework to guide students to think about the major ideas in the text. Teachers can also select questions to use as quizzes, rephrasing the questions into multiple choice or "true/false" formats. This provides assessment information about student understanding and serves as a motivational tool to ensure that students complete the reading assignment and comprehend the main ideas.

This is the stage of the activity that is most appropriate for teachers to explain concepts to students in whole-class lectures or discussions. References to **Blackline Masters** are available throughout the Teacher's Edition. They refer to illustrations from the textbook that teachers may photocopy and distribute to students or make overhead transparencies for lectures or presentations.

9. Review and Reflect

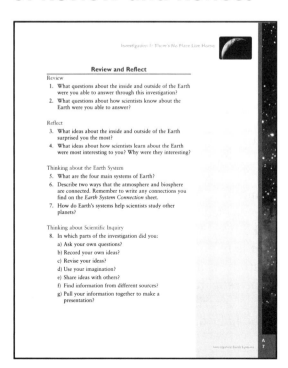

Questions in this feature ask students to use the key principles and concepts introduced in the activity. Students are sometimes presented with new situations in which they are asked to apply what they have learned. The questions in this section typically require higher-order thinking and reasoning skills than the **As You Read** questions. Teachers can assign these questions as homework, or have students complete them in groups during class. Assigning them as homework economizes time available in class, but has the drawback of making it difficult for students to collectively revisit the understanding that they developed as they worked through the concepts as a group during the investigation. A third alternative is, of course, to assign the work individually in class. When students work through application problems in class, teachers have the opportunity to interact with students at a critical juncture in their learning – when they may be just on the verge of "getting it."

Review and Reflect prompts students to think about what they have learned, how their work connects with the Earth system, and what they know about scientific inquiry. Another one of the important principles of learning used to guide the selection of content in *Investigating Earth Systems* was that:

"To develop competence in an area of inquiry, students must (a) have a deep foundation of factual knowledge, (b) understand facts and ideas in the context of a conceptual framework, and (c) organize knowledge in ways that facilitate retrieval and application." (*How People Learn: Bridging Research and Practice* National Research Council, 1999, P. 12.)

Reflecting on one's learning and one's thinking is an important metacognitive tool that makes students examine what they have learned in the activity and then think critically about the usefulness of the results of their inquiry. It requires students to take stock of their learning and evaluate whether or not they really understand "how it fits into the Big Picture." It is important for teachers to guide students through this process with questions such as "What part of your work demonstrates that you know and can do scientific inquiry? How does what you learned help you to better understand the Earth system? How does your work contribute or relate to the concepts of the Big Picture at the end of the module?"

10. Final Investigation: Putting It All Together

In the final investigation in each *Investigating Earth Systems* module, your students will apply all the knowledge they have about the topics explored to solve a practical problem or situation. Requiring students to apply all they have gained toward a specific outcome should serve as the main assessment information for the module. A sample assessment rubric is provided in the back of this Teacher's Edition. Whatever rubric you employ, it is important that you share this with students at the outset of the final investigation so that they understand the criteria upon which their work will be judged.

Investigating Earth Systems

The instructions provided to students are purposely open-ended, but can be completed to various levels depending upon how much knowledge students apply. During the final investigation, your role is to be a participant observer, moving from group to group, noticing how students go about the investigation and how they are applying the experience and understanding they have gained from the module.

11. Reflecting

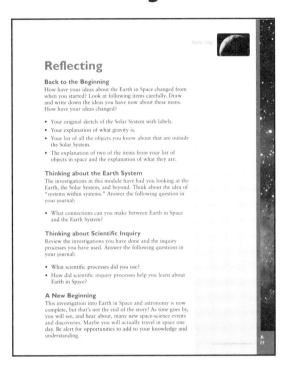

Now that students are at the end of the module, they are provided with questions that ask them to reflect upon all that they have learned about Earth science, inquiry, and the Earth system. The first set of questions (**Back to the Beginning**) are the same questions used in the pre-assessment. Teachers often ask students to revisit their initial responses and provide new answers to demonstrate how much they have learned.

12. The Big Picture

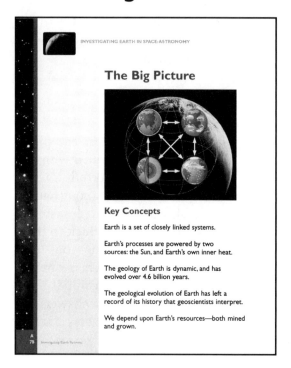

The five key concepts below underlie Earth science in general and *Investigating Earth Systems* in particular. Collectively, the ten modules in the *Investigating Earth Systems* series are designed to help students understand each of these concepts by the time they complete grade 8. Many of the concepts that underlie the Big Picture may be difficult for students to grasp easily. As students develop their ideas through inquiry-based investigations, you can help them to make connections with these key scientific concepts. As a reminder of the importance of the major understandings, the Student Book has a copy of the Big Picture in the back of the book near the **Glossary**.

Be on the lookout for chances to remind students that:
- Earth is a set of closely linked systems.
- Earth's processes are powered by two sources: the Sun and Earth's own inner heat.
- The geology of Earth is dynamic, and has evolved over 4.6 billion years.
- The geological evolution of Earth has left a record of its history that geoscientists interpret.
- We depend upon Earth's resources—both mined and grown.

Investigating Earth Systems

13. Glossary

Words that may be new or unfamiliar to students are defined and explained in the **Glossary** of the Student Book. Teachers use their own judgment about selecting the terms that appear in the **Glossary** that are most important for their students to learn. Teachers typically use discretion and consider their state and local guidelines for science content understanding when assigning importance to particular vocabulary, which in most cases is very likely to be a small subset of all the scientific terms introduced in each module and defined in the **Glossary**.

References

How People Learn: Bridging Research and Practice (1999) Suzanne Donovan, John Bransford, and James Pellegrino, editors. National Academy Press, Washington, DC. 78 pages. The report is also available online at www.nap.edu.

Using the Investigating Earth Systems Web Site

www.agiweb.org/ies

The *Investigating Earth Systems* web site has been designed for teachers and students.
- Each *Investigating Earth Systems* module has its own web page that has been designed specifically for the content addressed within that module.
- Module web sites are broken down by investigation and also contain a section with links to relevant resources that are useful for the module.
- Each investigation is divided into materials and supplies needed, **Background Information,** and links to resources that will help you and your students to complete the investigation.

Teacher's Edition

Enhancing Teacher Content Knowledge

Each *Investigating Earth Systems* module has a specific web page that will help teachers to gather further **Background Information** about the major topics covered in each activity.

Example from *Investigating Rocks and Landforms* – Investigation 1

Different Types of Rock

To learn more about different types of rocks, visit the following web sites:

• What are the basic types of rock?, Rogue Community College
This site lists the basic descriptions of sedimentary, metamorphic and igneous rocks. Detailed information on each type of rock is also available.
(http://www.jersey.uoregon.edu/~mstrick/AskGeoMan/geoQuerry13.html)

1. *Sedimentary Rocks:*
• Sedimentary Rocks, University of Houston
Detailed description of the composition, classification, and formation of sedimentary rocks.
(http://ucaswww.mcm.uc.edu/geology/maynard/INTERNETGUIDE/appendf.htm)
• Image Gallery for Geology, University of North Carolina
See more examples of sedimentary rocks.
(http://www.geosci.unc.edu/faculty/glazner/Images/SedRocks/SedRocks.html)
• Sedimentary Rocks Laboratory, Georgia Perimeter College
Read a thorough discussion of clastic, chemical, and organic sedimentary rocks. Illustrations accompany each description.
(http://www.gpc.peachnet.edu/~pgore/geology/historical_lab/sedrockslab.php)
• Textures and Structures of Sedimentary Rocks, Duke University
View a collection of slides of different sedimentary rocks as either outcrops or thin sections viewed through a microscope.
(http://www.geo.duke.edu/geo41/seds.htm)
• Sedimentary Rocks, Washington State University
Learn more about sedimentary processes, environments of deposition in relation to different sedimentary rocks. Topics covered include depositional environments, chemical or mechanical weathering, deposition and lithification, and classification.
(http://www.wsu.edu/~geology/geol101/sedimentary/seds.htm)

Obtaining Resources

The inquiry focus of *Investigating Earth Systems* will require teachers to obtain local or regional maps, rocks, and data. The *Investigating Earth Systems* web site helps teachers to find such materials. The web page for each *Investigating Earth Systems* module provides a list of relevant web sites, maps, videos, books, and magazines. Specific links to sources of these materials are often provided.

Managing Inquiry in Your Investigating Earth Systems Classroom

Materials

The proper management of materials can make the difference between a productive, positive investigation and a frustrating one. If your school has purchased the materials kit (available through It's About Time) most materials have been supplied. In many cases there will be additional items that you will need to supply as well. This can include photocopies or transparencies (**Blackline Masters** are available in the **Appendix**), or basic classroom supplies like an overhead projector or water source. On occasion, students will bring in materials. If you do not have the materials kit, a master list of materials for the entire module precedes the first investigation. Tips on using and managing materials accompany each investigation.

Safety

Each activity has icons noting safety concerns. In most cases, a well-managed classroom is the best preventive measure for avoiding danger and injury. Take time to explain your expectations before beginning the first investigation. Read through the investigations with your students and note any safety concerns. The activities were designed with safety in mind and have been tested in classrooms. Nevertheless, be alert and observant at all times. Often, the difference between an accident and a calamity is simple monitoring.

Time

This module can be completed in six weeks if you teach science in daily 45-minute class periods. However, there are many opportunities to extend the investigations, and perhaps to shorten others. The nature of the investigations allows for some flexibility.

An inquiry approach to science education requires the careful management of time for students to fully develop their investigative experience and skills. Most investigations will not easily fit into one 45-minute lesson. You may feel it necessary to extend them over two or more class periods. Some investigations include long-term studies. Where this is the case you may need to allow time for data collection each day, even after moving on to the next investigation.

Teacher's Edition

Classroom Space

On days when students work as groups, arrange your classroom furniture into small group areas. You may want to have two desk arrangements—one for group work and one for direct instruction or quiet work time.

The Student Journal

The student journal is an important component of each *IES* module. (See the **Appendix** in this Teacher's Edition for a **Blackline Master** of the Journal cover sheet.) Your students are young scientists as they investigate Earth science questions. Encourage your students to think of themselves in this way and to see their journals as records of their investigations.

The journal serves other functions as well. It is a key component in performance assessment, both formative and summative. (Formative evaluation involves the ongoing evaluation of students' level of understanding and their development of skills and attitudes. Summative evaluation is designed to determine the extent to which instructional objectives have been achieved for a topic.) Encourage your students to record observations, data, and experimental results in their journals. Answers to **Review and Reflect** questions at the end of each investigation should also be recorded in the journal. It is very important that students have enough time to review, reflect, and update their journals at the end of each investigation.

Frequent feedback is essential if students are to maintain good journals. This is difficult but not impossible. For many teachers, the prospect of collecting and grading anywhere from 20 to over 100 journals in a planning period, then returning them the next day, seems prohibitive. This does not need to be the case. If you use a simple rubric, and collect journals often, it is possible to grade 100 journals in an hour. It may not be necessary to write comments every time you collect journals; in some cases, it is equally effective to address trends in student work in front of the whole class. For example, students will inevitably turn in journals that contain no dates and/or headings. This leaves many questions unanswered and makes their work very hard to interpret. There is no need to write this comment over and over again! You might want to consider keeping your own teacher journal for this module. This makes a great template for evaluating student journals. In addition to documenting class activities, you might want to make notes on classroom management strategies, materials and supplies, and procedural modifications. Sample rubrics are included in the **Appendix**.

Student Collaboration

The National Science Education Standards and Benchmarks for Science Literacy emphasize the importance of student collaboration. Scientists and others frequently work in teams to investigate questions and solve problems. There are times, however, when it is important to work alone. You may have students who are more comfortable working this way. Traditionally, the competitive nature of school curricula has emphasized individual effort through grading, "honors" classes, and so on. Many parents will have been through this experience themselves as students and will be looking for comparisons between their children's performance and other students. Managing collaborative groups may therefore present some initial problems, especially if you have not organized your class in this way before.

Below are some key points to remember as you develop a group approach.

- Explain to students that they are going to work together. Explain *why* ("two heads are better than one" may be a cliché—but it is still relevant).
- Stress the responsibility each group member has to the others in the group.
- Choose student groups carefully to ensure each group has a balance of ability, special talents, gender and ethnicity.
- Make it clear that groups are not permanent and they may change occasionally.
- Help students see the benefits of learning with and from each other.
- Ensure that there are some opportunities for students to work alone (certain activities, writing for example, are more efficiently done in solitude).

Student Discussion

Encourage all students to participate in class discussions. Typically, several students dominate discussion while others hesitate to volunteer comments. Encourage active participation by explicitly stating that you value all students' comments. Reinforce this by not rejecting answers that appear wrong. Ask students to clarify contentious comments. If you ask for students' opinions, be prepared to accept them uncritically.

Teacher's Edition

Assessing Student Learning in Investigating Earth Systems

The completion of the final investigation serves as the primary source of summative assessment information. Traditional assessment strategies often give too much attention to the memorization of terms or the recall of information. As a result, they often fall short of providing information about students' ability to think and reason critically and apply information that they have learned. In *Investigating Earth Systems*, the solutions students provide to the final investigation in each module provide information used to assess thinking, reasoning, and problem-solving skills that are essential to life-long learning.

Assessment is one of the key areas that teachers need to be familiar with and understand when trying to envision implementing *Investigating Earth Systems*. In any curriculum model, the mode of instruction and the mode of assessment are connected. In the best scheme, instruction and assessment are aligned in both content and process. However, to the extent that one becomes an impediment to reform of the other, they can also be uncoupled. *Investigating Earth Systems* uses multiple assessment formats. Some are consistent with reform movements in science education that *Investigating Earth Systems* is designed to promote. **Project-based assessment,** for example, is built into every *Investigating Earth Systems* culminating investigation. At the same time, the developers acknowledge the need to support teachers whose classroom context does not allow them to depart completely from "traditional" assessment formats, such as paper and pencil tests.

In keeping with the discussion of assessment outlined in the National Science Education Standards (NSES), teachers must be careful while developing the specific expectations for each module. Four issues are of particular importance in that they may present somewhat new considerations for teachers and students. These four issues are dealt with on the next two pages.

Investigating Earth Systems: Earth in Space: Astronomy   xxxiii

1. Integrative Thinking

The National Science Education Standards (NSES) state: "Assessments must be consistent with the decisions they are designed to inform." This means that as a prerequisite to establishing expectations, teachers should consider the use of assessment information. In *Investigating Earth Systems*, students often must be able to articulate the connection between Earth science concepts and their own community. This means that they have to integrate traditional Earth science content with knowledge of their surroundings. It is likely that this kind of integration will be new to students, and that they will require some practice at accomplishing it. Assessment in one module can inform how the next module is approached so that the ability to apply Earth science concepts to local situations is enhanced on an ongoing basis.

2. Importance

An explicit focus of NSES is to promote a shift to deeper instruction on a smaller set of core science concepts and principles. Assessment can support or undermine that intent. It can support it by raising the priority of in-depth treatment of concepts, such as students evaluating the relevance of core concepts to their communities. Assessment can undermine a deep treatment of concepts by encouraging students to parrot back large bodies of knowledge-level facts that are not related to any specific context in particular. In short, by focusing on a few concepts and principles, deemed to be of particularly fundamental importance, assessment can help to overcome a bias toward superficial learning. For example, assessment of terminology that emphasizes deeper understanding of science is that which focuses on the use of terminology as a tool for communicating important ideas. Knowledge of terminology is not an end in itself. Teachers must be watchful that the focus remains on terminology in use, rather than on rote recall of definitions. This is an area that some students will find unusual if their prior science instruction has led them to rely largely on memorization skills for success.

3. Flexibility

Students differ in many ways. Assessment that calls on students to give thoughtful responses must allow for those differences. Some students will find the open-ended character of the *Investigating Earth Systems* module reports disquieting. They may ask many questions to try to find out exactly what the finished product should look like. Teachers will have to give a consistent and repeated message to those students, expressed in many different ways, that the ambiguity inherent in the open-ended character of the assessments is an opportunity for students to show what they know in a way that makes sense to them. This also allows for the assessments to be adapted to students with differing abilities and proficiencies.

4. Consistency

While the module reports are intended to be flexible, they are also intended to be consistent with the manner in which instruction happens, and the kinds of inferences that are going to be made about students' learning on the basis of them. The *Investigating Earth Systems* design is such that students have the opportunity to learn new material in a way that places it in context. Consistent with that, the module reports also call for the new material to be expressed in context. Traditional tests are less likely to allow this kind of expression, and are more likely to be inconsistent with the manner of teaching that *Investigating Earth Systems* is designed to promote. Likewise, in that *Investigating Earth Systems* is meant to help students relate Earth Science to their community, teachers will be using the module reports as the basis for inferences regarding the students' abilities to do that. The design of the module reports is intended to facilitate such inferences.

An assessment instrument can imply but not determine its own best use. This means that *Investigating Earth Systems* teachers can inadvertently assess module reports in ways that work against integrative thinking, a focus on important ideas, flexibility in approach, and consistency between assessment and the inferences made from that assessment.

All expectations should be communicated to students. Discussing the grading criteria and creating a general rubric are critical to student success. Better still, teachers can engage students in modifying and/or creating the criteria that will be used to assess their performance. Start by sharing the sample rubric with students and holding a class discussion. Questions that can be used to focus the discussion include: Why are these criteria included? Which activities will help you to meet these expectations? How much is required? What does an "A" presentation or report look like? The criteria should be revisited throughout the completion of the module, but for now students will have a clearer understanding of the challenge and the expectations they should set for themselves.

Investigating Earth Systems Assessment Tools

Investigating Earth Systems provides you with a variety of tools that you can use to assess student progress in concept development and inquiry skills. The series of evaluation sheets and scoring rubrics provided in the back of this Teacher's Edition should be modified to suit your needs. Once you have settled on the performance levels and criteria and modified them to suit your particular needs, make the evaluation sheets available to students, preferably before they begin their first investigation. Consider photocopying a set of the sheets for each student to include in his or her journal. You can also encourage your students to develop their own rubrics. The final investigation is well-suited for such, since students will have gained valuable experience with criteria by the time they get to this point in the module. Distributing and discussing the evaluation sheets will help students to become familiar with and know the criteria and expectations for their work. If students have a complete set of the evaluation sheets, you can refer to the relevant evaluation sheet at the appropriate point within an *IES* lesson.

1. Pre-Assessment

The pre-assessment activity culminates with students putting their journals together and adding their first journal entry. It is important that this not be graded for content. Credit should be given to all students who make a reasonable attempt to complete the activity. The purpose of this pre-assessment is to provide a benchmark for comparison with later work. At the end of the module, the central questions of the pre-assessment are repeated in the section called **Back to the Beginning**.

Teacher's Edition

2. Assessing the Student Journal

As students complete each investigation, reinforce the need for all observations and data to be organized well and added to the journals. Stress the need for clarity, accurate labeling, dating, and inclusion of all pertinent information. It is important that you assess journals regularly. Students will be more likely to take their journals seriously if you respond to their work. This does not have to be particularly time-consuming. Five types of evaluation instruments for assessing journal entries are available at the back of this Teacher's Edition to help you provide prompt and effective feedback. Each one is explained in turn below.

Journal Entry-Evaluation Sheet

This sheet provides you with general guidelines for assessing student journals. Adapt this sheet so that it is appropriate for your classroom. The journal entry evaluation sheet should be given to students early in the module, discussed with students, and used to provide clear and prompt feedback.

Journal Entry-Checklist

This checklist provides you and your students with a guide for quickly checking the quality and completeness of journal entries. You can assign a value to each criterion, or assign a "+" or "-" for each category, which you can translate into points later. However you choose to do this, the point is to make it easy to respond to students' work quickly and efficiently. Lengthy comments may not be necessary. Depending on time constraints, you may not have time to write comments each time you evaluate journals. The important thing is that students get feedback—they will do better work if they see that you are monitoring their progress.

Key Question-Evaluation Sheet

This sheet will help students to learn the basic expectations for the warm-up activity. The **Key Question** is intended to reveal students' conceptions about the phenomena or processes explored in the activity. It is not intended to produce closure, so your assessment of student responses should not be driven by a concern for correctness. Instead, the evaluation sheet emphasizes that you want to see evidence of prior knowledge and that students should communicate their thinking clearly. It is unlikely that you will have time to apply this assessment every time students complete a warm-up activity, yet in order to ensure that students value committing their initial conceptions to paper and taking the warm-up seriously, you should always remind students of the criteria. When time permits, use this evaluation sheet as a spot check on the quality of their work.

Investigation Journal Entry-Evaluation Sheet

This sheet will help students to learn the basic expectations for journal entries that feature the write-up of investigations. *IES* investigations are intended to help students to develop content understanding and inquiry abilities. This evaluation sheet provides a variety of criteria that students can use to ensure that their work meets the highest possible standards and expectations. When assessing student investigations, keep in mind that the **Investigate** section of an *IES* lesson corresponds to the explore phase of the learning cycle (engage, explore, apply, evaluate) in which students explore their conceptions of phenomena through hands-on activity. Using and discussing the evaluation sheet will help your students to internalize the criteria for their performance. You can further encourage students to internalize the criteria by making the criteria part of your "assessment conversations" with them as you circulate around the classroom and discuss student work. For example, while students are working, you can ask them criteria-driven questions such as: "Is your work thorough and complete? Are all of you participating in the activity? Do you each have a role to play in solving the problem?" and so on.

Review and Reflect Journal Entry-Evaluation Sheet

Reviewing and reflecting upon one's work is an important part of scientific inquiry and is also important to learning science. Depending upon whether you have students complete the work individually or within a group, the **Review and Reflect** portion of each investigation can be used to provide information about individual or collective understandings about the concepts and inquiry processes explored in the investigation. Whatever choice you make, this evaluation sheet provides you with a few general criteria for assessing content and thoroughness of student work. Adapt and modify the sheet to meet your needs. Consider involving students in selecting and modifying the criteria for evaluating their end of investigation reflections.

3. Assessing Group Participation

One of the challenges to assessing students who work in collaborative teams is assessing group participation. Students need to know that each group member must pull his or her weight. As a component of a complete assessment system, especially in a collaborative learning environment, it is often helpful to engage students in a self-assessment of their participation in a group. Knowing that their contributions to the group will be evaluated provides an additional motivational tool to keep students constructively engaged. These evaluation forms (Group Participation Evaluation Sheets I and II) provide students with an opportunity to assess group participation. In no case should the results of this evaluation be used as the sole source of assessment data. Rather, it is better to assign a weight to the results of this evaluation and factor it in with other sources of assessment data. If you have not done this before, you may be surprised to find how honestly students will critique their own work, often more intensely than you might do.

4. Assessing the Final Investigation

Students' work throughout the module culminates with the final investigation. To complete it, students need a working knowledge of previous activities. Because it refers back to the previous steps, the last investigation is a good review and a chance to demonstrate proficiency. For an idea on how to use the last investigation as a performance-based exam, see the section in the **Appendices**.

5. Assessing Inquiry Processes

There is an obvious difficulty in assessing individual student proficiency when the students work within a collaborative group. One way to do this is to have a group present its results followed by a question-and-answer session. You can direct questions to individual students as a way of checking proficiency. Another is to have every student write a report on his or her role in the investigation, after first making it clear what this report should contain. Individual interviews are clearly the best option but may not be feasible given the time constraints of most classes.

6. Traditional Assessment Options

A traditional paper-and pencil-exam is included in the **Appendices**. While performance-based assessments may offer teachers more insight into student skill levels, computer-generated tests are also useful—especially so in states with state-sponsored exams. Additionally, some students are strong in one area and not as strong in another. Using a variety of methods for assessing and grading students' progress offers a more complete picture of the success of the student—and the teacher.

Reviewing and Reflecting upon Your Teaching

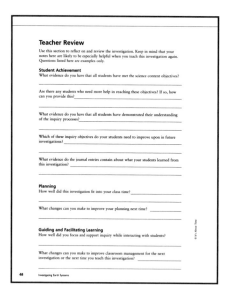 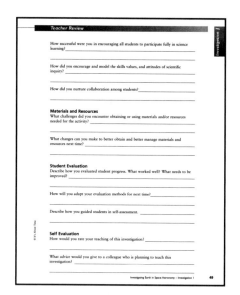

Reviewing and Reflecting upon Your Teaching provides an important opportunity for professional growth. A two-page Teacher Review form is included at the end of each investigation. The purpose of these reviews is to help you to reflect on your teaching of each investigation. We suggest that you try to answer each question at the completion of each investigation, then go back to the relevant section of this Teacher's Edition and write specific comments in the margins. Use the comments the next time you teach the investigation. For example, if you found that you were able to make substitutions to the list of materials needed, write a note about those changes in the margin of that page of this Teacher's Edition.

NOTES

NOTES

Investigating Earth in Space: Astronomy
Introduction

Stargazing has been going on since the very earliest times, though how the first humans interpreted the heavens is hard to imagine. They would have noticed the regular patterns of what we now call constellations, and would have become familiar with how these "moved" during the night and perhaps the pattern of change taking place over a year. They may well have attributed mystical ideas and religious significance to the heavenly bodies above.

It is not until about 2354 B.C. through the poems of En Hedu'anna, chief priestess of the Moon Goddess, that we have any evidence of how early peoples viewed the night sky. The daughter of Sargon (of Akkad), who established the Sargonian Dynasty in Babylon, En Hedu'anna held a position of great power and prestige. We now know that the priests and priestesses of Babylon and Sumeria established a network of observatories to monitor the movements of the stars as early as 3000 B.C. These represent the origins of astronomy, as well as mathematics, which follow a common line of development from then to the present. Indeed, certain religious events like Easter and Passover still follow the calendar they created.

From here the development of the science of Astronomy passes to the Greeks. Starting about 600 B.C. with Thales through Pythagoras, Socrates, Democritus, Plato, Aristotle, Erastosthenes, and Hipparachus, to Ptolemy in about A.D. 165 Astronomy, for the Greeks, was mainly tied to keeping time. Their love of mathematics and philosophy stretched human understanding of the universe and introduced the idea of a calendar. The idea that the Earth was a sphere, that the orbit of the Moon was inclined to the Equator of the Earth, and that Venus as an evening "star" was the same planet as Venus as a morning star were important discoveries.

The next major period of development came about 1000 years later in Western Europe from about 1450 – 1650. This period included such figures as: Copernicus, in Poland; Tycho Brahe, in Denmark; Johannes Kepler, and Maria Kirch in Germany; Galileo Galilei and, to some extent, Leonardo Da Vinci, in Italy; and Isaac Newton and Edmund Halley in England. They all played key roles in shaping the science and laws of astronomy as we know it today.

From the 18th century onward, distinguished scientists from all over the world, have continued to develop the science of astronomy. These people include: Caroline Herschel, from Germany and England (1750-1848); Maria Mitchell, U.S.A. (1818-1889); and more recently Evelyn Granville (1924 -) who has made huge contributions to the Space Shuttle program; and Jocelyn Bell Burnell from England, who discovered Pulsars in 1968. Other well-known scientists include: Robert Goddard, William Fleming, Albert Einstein, George Hale, Edwin Hubble, Carl Sagan and, more recently, Stephen Hawkins. Their work and discoveries, along with those of many others, make up the exciting story of astronomy.

Today, many of those lights in the night sky that so mystified our ancestors have become part of our daily lives. A new era of space investigation began with the launch of Sputnik, the Russian artificial satellite, in 1957. This prompted the U.S. into rapid development of its own space aspirations, and the "Space Race" was on. Russia continued her early success by being the first to send an object to the Moon. Russia then launched Yuri Gagarin, the first person, into space in 1961 followed by the first woman, Valentina Tereskhkova, in 1963. Spurred on by these successes, the U.S. mounted a huge effort with the missions of the Gemini program. In just 20 short months, between March 1965 and November 1966, 10 Gemini crews pioneered the techniques necessary for a lunar mission. Fewer than three years later, the Apollo 11 mission landed Neil Armstrong and Buzz Aldrin on the surface of the Moon. Of equal importance was the fact that most of this was watched by millions on television. Space science had truly come of age and was now part of popular experience.

Many of the technological advances that made these space travel achievements possible moved into everyday use. These included computers, automated appliances, communication technology, transportation, and many others. Sights were now set on other bodies in the Solar System and beyond. Successful robotic missions to Mars, Saturn, and Jupiter have virtually taken us to places we could only see as science fiction fifty years ago. Space satellites and sensors are providing us with huge amounts of as-it-happens information. We now have amazing tools with which to explore the farthest reaches of the universe.

Modern middle school students will become adults in a technical world far different from that of their parents and grandparents. They will witness the "opening up" of space in a form that is hard to imagine today. In a world where rapid technological change will be the norm and space exploration an important driver, a good understanding of Earth and space will be critical to enable citizens to make sound political and economic judgments, and make the most of future opportunities.

More Information...on the Web

Visit the *Investigating Earth Systems* web site www.agiweb.org/ies/ for links to a variety of other web sites that will help you deepen your understanding of content and prepare you to teach this module, *Investigating Earth in Space: Astronomy*.

Introduction

Students' Conceptions about Earth and Space

It would be wonderful if every student had, at some point, the experience of lying in a field at night observing the heavens above, seeing a "shooting star," and watching the arrival of dawn on the horizon or the Sun setting below it. But you can't count on this.

The ideas that students hold about Earth and space depend upon experiences they have had. If their elementary school curriculum included Earth and space science, they should know about objects in the sky. They should know that the Sun, Moon, and stars have properties, locations, and movements that can be observed (along with clouds, birds, and airplanes.) They are likely to understand that the Sun provides enough light and heat to maintain the temperature of the Earth. They will know that the Sun appears to move across the sky in more or less the same way every day, but that its path changes slowly over the seasons. They will probably know that the Moon moves across the sky on a daily basis much like the Sun, and that the observable shape of the Moon changes from day to day in a cycle that lasts about a month.

If students have not covered these topics in elementary school, they may have acquired these understandings from alternative sources such as family and friends, scouting groups, science center activities, television science shows, movies, books, informal education programs, or from their own observations and questions. However, it cannot be assumed that all students have had these experiences or the opportunity to reflect upon appropriate observations and ideas. Students living in dense urban settings may less often see a wide expanse of sky compared to their suburban and rural counterparts and, for them, observations of the night sky may be diluted by street lamps and other light "pollution."

When it comes to the Solar System, middle school students may have a variety of informal ideas that do not conform to commonly accepted scientific explanations. For example, some may think that the Earth is the largest object and also the center of the Solar System. They may make little distinction between the Solar System, the Milky Way Galaxy, and the universe, thinking stars and planets are the same thing, that all stars are the same size, and that they are evenly distributed throughout the universe. In fact, the only thing that can be counted upon, in an average collection of middle school students, is that ideas about Earth and space will comprise a wide range of belief and knowledge. A study by the American Institute of Physics identified a number of popular misconceptions that students can have, which included:

Ideas about Astronomy
- Stars and constellations appear in the same place in the sky every night.
- The Sun rises exactly in the east and sets exactly in the west every day.
- We experience seasons because of the Earth's changing distance from the Sun (closer in the summer, farther in the winter).
- The Moon can only be seen during the night.
- The Moon does not rotate on its axis as it revolves around the Earth.
- The phases of the Moon are caused by shadows cast on its surface by other objects in the Solar System.

- The phases of the Moon are caused by the shadow of the Earth on the Moon.
- The phases of the Moon are caused by the Moon moving into the Sun's shadow.
- The shape of the Moon always appears the same.
- The Earth is the largest object in the Solar System.
- Meteors are falling stars.
- Comets and meteors are out in space and do not reach the ground.
- The surface of the Sun is without visible features.
- All the stars in a constellation are near each other.
- All the stars are the same distance from the Earth.
- The brightness of a star depends only on its distance from the Earth.
- Stars are evenly distributed throughout our Galaxy.

Ideas about Space
- The Earth is sitting on something.
- The Earth is larger than the Sun.
- The Sun disappears at night.
- The Earth is round like a pancake.
- We live on the flat middle of a sphere.
- There is a definite up and down in space.
- Seasons are caused by the Earth's distance from the Sun.
- Phases of the Moon are caused by a shadow from the Earth.
- Different countries see different phases of the Moon on the same day.
- The amount of daylight increases each day of summer.
- Planets cannot be seen with the naked eye.
- Planets appear in the sky in the same place every night.
- Astrology is able to predict the future.
- Gravity is selective; it acts differently or not at all on some matter.
- Gravity increases with height.
- Gravity requires a medium to act through.
- Rockets in space require a constant force.
- The Sun will never burn out.
- The Sun is not a star.

Introduction

NOTES

Investigating Earth in Space: Astronomy
Module Flow

Activity Summaries	Emphasis
Pre-Assessment Students describe their understanding of key concepts explored in the module.	Recording initial knowledge and understanding of content.
Introducing Earth in Space: Astronomy Students discuss their ideas and experiences related to the topics they will be investigating.	Putting the investigations into a meaningful context.
Investigation 1: There's No Place Like Home Students make drawings of what they think Earth looks like from space and as a cross-section, share results, and compare them to accurate diagrams.	Modeling, making observations and inferences. Raising questions through using models, recording observations, consulting information sources and sharing findings.
Investigation 2: The Earth's Moon Students work through stations to learn about the characteristics of the Earth's Moon. They build a model to investigate their ideas.	Modeling, making observations. Presenting to and sharing findings with others.
Investigation 3: The Earth and Its Moon Students work through stations to learn about the characteristics of the Earth's Moon and engage in a series of activities to understand gravity.	Using a variety of resources to find credible scientific information and making a scale model to demonstrate a scientific concept.
Investigation 4: Finding Our Place in Space Students use a variety of resources to learn about where the Earth is in space. They then make a model of the Solar System that they can display in the school's hallways.	Investigating natural processes using models, conducting tests and drawing conclusions.
Investigation 5: The Sun and Its Central Role in our Solar System Students compare their ideas to established knowledge about the Sun, then conduct investigations to understand the Sun's energy and its importance to the Earth.	Comparing current ideas with accepted scientific explanations, conducting tests, and drawing conclusions.
Investigation 6: The Planetary Council Student groups specialize in learning about a particular planet so that they can defend why their planet should receive federal funding to be studied.	Using a variety of resource formats to collect accurate scientific information. Organizing information into a proposal that others can see and understand.
Investigation 7: What is Beyond Our Solar System? Students conduct a series of investigations to learn about the light from stars, as well as about constellations, galaxies and nebulae.	Conducting experiments to reveal scientific phenomena and relating this to the bigger picture of the Solar System and beyond.
Investigation 8: Discovering the Difference Between Science Fact and Science Fiction Students are asked to distinguish between science fact and fiction by producing a science fiction communication piece.	Reviewing the content and inquiry processes that have been used throughout the module.
Reflecting Students review the science content and inquiry processes they used throughout the module.	Assessing student learning.

Introduction

Investigating Earth in Space: Astronomy
Module Objectives

Investigation	Science Content	Inquiry Process Skills
Investigation 1: There's No Place Like Home Students investigate what Earth looks like from space and within its interior.	Students will collect evidence that: 1. Explores how the Earth looks from space. 2. Explores what the interior of the Earth is like.	Students will: 1. Make observations using their senses. 2. Record observations in a systematic way. 3. Make a model of what they cannot see. 4. Revise models based on additional observations. 5. Communicate observations and findings to others.
Investigation 2: The Earth's Moon Students work through stations to learn about characteristics of Earth's Moon.	Students will collect evidence that: 1. Demonstrates the surface of the Moon has craters. 2. The Moon's appearance goes through phases.	Students will: 1. Investigate using a variety of information and data resources. 2. Build a model to investigate ideas. 3. Collect and record data. 4. Communicate observations and findings to others.
Investigation 3: The Earth and Its Moon Students explore the relationship between the Earth and its Moon. Then they engage in activities to understand the nature of gravity.	Students will collect evidence that: 1. Helps them define gravity. 2. The Moon affects Earth's tides. 3. Explores how gravity works in the Solar System.	Students will: 1. Investigate using a variety of information and data resources. 2. Use models and tools to reveal ideas and information. 3. Make observations and record results. 4. Use data tables to investigate phenomena 5. Communicate observations and findings to others.

Investigating Earth in Space: Astronomy
Module Objectives

Investigation	Science Content	Inquiry Process Skills
Investigation 4: Finding Our Place in Space Students use a variety of resources to learn about where the Earth is in space. They then make a model of the Solar System that they can display in the school's hallways.	Students will collect evidence that: 1. The Earth is the third planet from the Sun. 2. Some planets are larger than the Earth; others are smaller. 3. The planets all orbit the Sun and are different distances from the Sun and each other.	Students will: 1. Use a variety of investigative resources. 2. Make a model of what they cannot see. 3. Revise models on the basis of additional observations. 4. Communicate observations and findings to others.
Investigation 5: The Sun and Its Central Role in Our Solar System Students compare their ideas to established knowledge. about the Sun, then conduct investigations to understand the Sun's energy and its importance to the Earth.	Students will collect evidence that: 1. The Earth receives energy from the Sun. 2. Life on Earth is dependent upon energy from the Sun. 3. The Sun is a star — a ball of burning gases.	Students will: 1. Compare their own scientific explanations with established scientific explanations. 2. Use models to investigate science questions. 3. Collate information into a useful format. 4. Communicate observations and findings to others.
Investigation 6: The Planetary Council Student groups specialize in learning about a particular planet so that they can defend why their planet should receive federal funding to be studied.	Students will collect evidence that: 1. A planet has unique characteristics. 2. The remaining planets in the Solar System have unique characteristics.	Students will: 1. Locate and review relevant information resources. 2. Identify key scientific features relating to an object. 3. Communicate observations and findings to others.

Introduction

Investigating Earth in Space: Astronomy
Module Objectives

Investigation	Science Content	Inquiry Process Skills
Investigation 7: What is Beyond Our Solar System? Students conduct a series of investigations to learn about the light from stars, as well as about constellations, galaxies, and nebulae.	Students will collect evidence that: 1. Very bright light that is far away from the eye can appear the same as dimmer light that is closer to the eye. 2. Space contains many objects, including galaxies and nebulae. 3. Our Solar System is part of the Milky Way Galaxy.	Students will: 1. Conduct investigations that model scientific processes. 2. Use the findings to interpret scientific phenomena.
Investigation 8: Discovering the Difference Between Science Fact and Science Fiction In this final investigation, students are asked to distinguish between science fact and fiction and produce a science fiction communication piece.	Students will demonstrate that they have: 1. A clear understanding of astronomical concepts and processes. 2. An ability to distinguish between science fact and science fiction based on evidence. 3. An ability to weave science fact and science fiction together to produce a convincing communications piece.	Students will: 1. Conduct investigations that model scientific processes. 2. Collect and review both scientific information and science fiction techniques. 3. Work as a team to create and design a captivating communication piece. 4. Distinguish between commonly accepted scientific concepts and practices and pseudo-scientific fiction ideas.

National Science Education Content Standards

Investigating Earth Systems is a standards-driven curriculum. That is, the scope and sequence of the series is derived from, and driven by, the National Science Education Standards (NSES) and the American Association for the Advancement of Science (AAAS) Benchmarks for Science Literacy (BSL). Both specify content standards that students should know by the completion of eighth grade.

Unifying Concepts and Processes
- Systems, order, and organization
- Evidence, models, and explanation
- Constancy, change, and measurement
- Evolution and equilibrium

Science as Inquiry
- Identify questions that can be answered through scientific investigations
- Design and conduct a scientific investigation
- Use tools and techniques to gather, analyze, and interpret data
- Develop descriptions, explanations, predictions, and models based upon evidence
- Think critically and logically to make the relationships between evidence and explanation
- Recognize and analyze alternative explanations and predictions
- Communicate scientific procedures and explanations
- Use mathematics in all aspects of scientific inquiry
- Understand scientific inquiry

Physical Science
- Properties and changes of properties in matter
- Motion and forces
- Transfer of energy

Life Science
- Diversity and adaptation of organisms

Earth and Space Science
- Structure of the Earth system
- Earth's history
- Earth in the Solar System

Science and Technology
- Abilities of technological design
- Understandings about science and technology

Science in Personal and Social Perspectives
- Natural hazards
- Risks and benefits
- Science and technology in society

History and Nature of Science
- Science as a human endeavor
- Nature of science
- History of science

Introduction

Key NSES Earth Science Standards Addressed in IES Astronomy

1. The Earth is the third planet from the Sun in a system that includes the Moon, the Sun, eight other planets and their moons, and smaller objects, such as asteroids and comets. The Sun, an average star, is the central and largest body in the Solar System.

2. Most objects in the Solar System are in regular and predictable motion. Those motions explain such phenomena as the day, the year, phases of the Moon, and eclipses.

3. Gravity is the force that keeps planets in orbit around the Sun and governs the rest of the motion in the Solar System. Gravity alone holds us to the Earth's surface and explains the phenomena of the tides.

4. The Sun is the major source of energy for phenomena on the Earth's surface, such as growth of plants, winds, ocean currents, and the water cycle. Seasons result from variations in the amount of the Sun's energy hitting the surface, due to the tilt of the Earth's rotation on its axis and the length of the day.

Understandings About Science and Technology

1. Scientific inquiry and technological design have similarities and differences. Scientists propose explanations for questions about the natural world, and engineers propose solutions relating to human problems, needs, and aspirations. Technological solutions are temporary; technologies exist within nature and so they cannot contravene physical or biological principles; technological solutions have side effects; and technologies cost, carry risks, and provide benefits.
[See Content Standards A, F, & G (grades 5–8)]

2. Many different people in different cultures have made and continue to make contributions to science and technology.

3. Science and technology are reciprocal. Science helps drive technology, as it addresses questions that demand more sophisticated instruments and provides principles for better instrumentation and technique. Technology is essential to science, because it provides instruments and techniques that enable observations of objects and phenomena that are otherwise unobservable due to factors such as quantity, distance, location, size, and speed. Technology also provides tools for investigations, inquiry, and analysis.

4. Perfectly designed solutions do not exist. All technological solutions have trade-offs, such as safety, cost, efficiency, and appearance. Engineers often build in back-up systems to provide safety. Risk is part of living in a highly technological world. Reducing risk often results in new technology.

5. Technological designs have constraints. Some constraints are unavoidable, for example, properties of materials, or effects of weather and friction; other constraints limit choices in the design, for example, environmental protection, human safety, and aesthetics.

6. Technological solutions have intended benefits and unintended consequences. Some consequences can be predicted, others cannot.

NOTES

Introduction

Materials and Equipment List for Investigating Earth in Space: Astronomy

Pre-Assessment
Each group of students will need:
- poster board, poster paper, or butcher paper
- Student Journal cover sheet, one for each student: (**Blackline Master Astronomy P. 2,** available at the back of this Teacher's Edition)

Teachers will need:
- overhead projector, chalkboard, or flip-chart paper
- transparency of **Blackline Master Astronomy P. 1** (Questions about Astronomy)

Materials Needed for Each Group per Investigation
Note: The *Investigating Earth Systems* web site www.agiweb.org/ies/ provides links to topical Internet sites and a list of resources that will aid student research.

Investigation 1
- colored pencils or markers and blank paper
- relief map or globe of the Earth
- 5" x 8" index cards
- computer with Internet access, if possible

Investigation 2
- Moon maps
- flashlights or overhead projector light
- Tennis ball
- Ping-Pong™ ball
- chart of Moon phases
- learning station materials (poster board, markers, blank paper, tape, stapler)
- metric ruler
- D-cell battery
- masking tape
- colored pencils
- 5" x 8" index cards
- flour
- poster board

Investigation 3
- metric measuring tape
- calculators
- masking tape

Investigation 4
- laminated diagram of the Solar System
- Solar System Poster
- calculators
- safety scissors
- colored pencils
- roll of adding machine tape
- masking tape

Investigation 5
- triangular prism
- thermometers
- colored pencils
- poster board
- LED light
- Plant that has been left in a closet for 1 week
- Plant that has been left in sunlight

Investigation 6
- colored pencils
- poster board

Investigation 7
- measuring tape
- flashlight
- star chart book
- LED light
- pack of glow-in the dark stars
- black poster board

Investigation 8
- colored pencils
- poster board

Introduction

General Supplies

Although the investigations can be done with the specific materials listed, it is always a good idea to build up a supply of general materials.

- 2 or 3 large clear plastic storage bins about 30 cm x 45 cm x 30 cm deep, with lids (these can be used for storage and also make good water containers)
- 2 or 3 plastic buckets and one large water container (camping type with a faucet)
- rolls of masking tape, duct tape, and clear adhesive tape
- rolls of plastic wrap and aluminum foil
- clear self-locking plastic bags (various sizes)
- ball of string and spools of sewing thread
- pieces of wire (can be pieces of wire coat hangers)
- stapler, staples, paper clips, and binder fasteners
- safety scissors and one sharp knife
- cotton balls, tongue depressors
- plastic and paper cups of various sizes
- empty coffee and soup cans, empty boxes and egg cartons
- several clear plastic soda bottles (various sizes)
- poster board, overhead transparencies, tracing paper, and graph paper
- balances and/or scales, weights, spring scales
- graduated cylinders, hot plates, microscopes
- safety goggles
- lab aprons or old shirts
- first aid kit

Teaching the Nature of Science

Students need meaningful experiences in which they engage in scientific investigations to fully understand the how and why of conducting inquiry in science. These experiences provide the foundation for and awareness of the history of science. Historical examples demonstrate that science is a human endeavor and show how people of different backgrounds have in the past and continue to contribute to scientific discovery.

A major distinction between science and other ways of knowing and bodies of knowledge is that science strives for the best and simplest explanation of the natural world through the use of logical argument, empirical standards, and skepticism. These explanations must meet certain criteria. Explanations based on personal beliefs, myths, mystical inspiration, religious values, or superstitions may have personal value but they have no place in scientific thought. Scientific explanations must be consistent with observations and experimental evidence about nature, must make accurate predictions, must be logical, must be open to criticism, and must be adaptable to change. These explanations and knowledge gained must also be shared with the general public in an appropriate manner.

Science is one of the few endeavors that constantly attempts to make its current understanding and interpretations obsolete. Imagine a researcher who discovered a cure for a certain disease. This scientist's success has rendered the work by others researching this disease as well as the doctors treating the disease obsolete. In fact, the success eliminated the researcher's own job. How many other disciplines can make this statement?

Many of your students may recognize that the various science disciplines differ in their topics, techniques, and outcomes. For instance, a naked-eye observational study in zoology can yield a plethora of data; whereas this same technique applied to quantum mechanics provides little, if any, meaningful information. While noting these differences, you should also highlight the commonalities of purpose, philosophy, and enterprise across these science disciplines. The arbitrary and fictitious lines drawn separating the sciences lead many students to false conclusions. Many students fail to see how biologists depend on chemistry, physics, Earth science, mathematics, and communication skills to fully understand and explain their discipline.

Much of the history of science has focused on the European discoveries. Most students recognize the names and accomplishments of Galileo, Newton, and Kepler to name a few. Non-European cultures have developed scientific ideas and solved problems through the application of technology. Individuals of diverse backgrounds, interests, talents, and motivations have made many contributions. The Far Eastern and other ancient cultures have provided much to build the body of science but often these contributors go unmentioned. Studying some of these individuals, as well as those who are active today, provides your students with further understanding of scientific inquiry and science as a human endeavor. Presenting a "world-view" of those participating in past and present scientific activities enables *all* of your diverse students to become a connected, dynamic member of the group known as the "inquirers of nature."

A common thought held by many students is that people from the past, especially historical figures, are inferior to today's people because they did not know what we

Introduction

know today. This is particularly acute for those historical figures whose theories have been displaced. You might have your students project how naïve their own thoughts about how the world works today would be compared to students' views in the year 4000. Emphasize that having a theory found to be incorrect does not make the presenter of the theory inferior. New techniques, tools, and models lead to better predictions and explanations.

Scientific discoveries mean very little if they are not shared with others. No matter who does science and mathematics or invents things, when or where they do it, the knowledge and technology that results can become available to everyone. The United States space program of the 1960s resulted in a wealth of products now used by most members of our society. Powdered orange juice, hand-held calculators, solar collectors, Velcro®, and even a ballpoint pen that writes at any angle are now ingrained in our culture, yet all had their origin in the Apollo missions to the moon. Kevlar® bullet-proof vests, GPS navigation systems, and fingerprint identification systems to access your personal computer files are but a few of the more recent scientific discoveries that have made their way to mainstream America. Who knows what future products now reside on a researcher's workbench awaiting release to the world?

Awaiting release to the world…sounds a bit ominous. One important responsibility of a scientist is to be a good steward to our world. This implies that a scientist must not release anything to the environment that is harmful. Additionally, science ethics demand that scientists must not knowingly subject coworkers, students, the neighborhood, or the community to health or property risks. Students may rightly become quite confused over the previous statement. A researcher developing military applications could be viewed by many as contributing to health or property risks of others; however, the discussion can lead to the protective benefits of having weapon systems. Having a weapons system can, in many cases, actually reduce the likelihood that these systems will be used.

Some experimental studies use human subjects. These studies are necessary to test the effectiveness and unexpected effects prior to the release of treatment options. In research involving human subjects, the ethics of science require the potential subjects be fully informed about the risks and benefits associated with the research and of their right to refuse to participate. All human subject investigations are voluntary in nature. No researcher can force a person to participate or continue in a study without the consent of the person. This was not always the case. In earlier times prisoners and others were forced to participate in studies or were not informed and took part in testing without their knowledge. Serious medical and ethical implications led to the current, strict guidelines for human subject testing.

Whereas a human can deny or withdraw consent to take part in an experiment, animals used in testing have no such option. It is therefore up to review boards and cultural norms to establish guidelines for the necessity and requirements for testing involving animal subjects. A strong needs statement must accompany all requests for animal testing. Without a powerful rationale for the inclusion of animals testing, the request is in many cases denied. Care prior to, during, and post treatment must be provided for all animals used in investigations. You should remind your students that animals used in observational studies in the classroom are subject to the same care rules as those used in more traditional investigations.

Pre-assessment

Overview

During the pre-assessment phase, the students complete an open-ended survey of their knowledge and understanding of key concepts explored in the Astronomy module. Students are given four items to consider and their responses become the first entry in their journals.

Preparation and Materials Needed

This pre-assessment activity does not appear in the student book. Yet it is crucial that you conduct this pre-assessment before introducing this module to your students and distributing the student books. When you complete the pre-assessment activity, you will have important data that tell you what your students already know about astronomy.

The pre-assessment should not be presented as a test. Make sure that your students are clear about this. Tell students that, at the end of their investigations, they will be able to compare how their ideas and knowledge about astronomy have changed as a result of their investigations, and that you will also be able to gauge how successful the investigations have been for everyone.

After the pre-assessment, and before distributing the student books, take some time to reflect on the ideas your students have. This is the starting point. You need to ensure that what follows fits with your students' prior knowledge.

Materials:
- poster board, poster paper, or butcher paper
- overhead projector, chalkboard, or flip-chart paper
- overhead transparency of questions (**Blackline Master Astronomy P. 1**)
- student journal cover sheet, one for each student (**Blackline Master Astronomy P. 2**)

Suggested Teaching Procedure

1. Let students know that what they write in this exercise will become their first entry in a journal that they will keep throughout the module. Explain that each person is going to write down ideas that they have about astronomy. The reason for this is to provide them and you with a starting point for their investigations into astronomy. Tell students that when they have finished the module, they will respond to these same items again. This will allow them and you to compare how their knowledge about astronomy has changed as a result of their investigations.

Introduction

2. Display the pre-assessment questions on an overhead projector, or provide each student with a copy of the questions (**Blackline Master Astronomy P. 1**). Have students write their responses in their journals.

Allow a reasonable period of time for all students to respond. Circulate around the classroom, prompting students to provide as much detail as possible.

> - Draw a sketch of the Solar System and label as many objects as you can in it.
> - Explain what gravity is.
> - Make a list of all the objects you know about in space that are outside the Solar System.
> - Choose two items from your list of objects in space and explain what they are.

3. Give each student a copy of the journal cover sheet (**Blackline Master Astronomy P. 2**). Direct students to insert the journal cover sheet and their pre-assessment into their journal. Explain that they now have one of the most important tools for this investigation into astronomy: their own journal.

Teaching Tip

What form will journals take? Using loose-leaf notebook paper in a thin three-ring binder enables students to add or remove pages easily. On the downside, loose-leaf pages are more easily lost and students must maintain a regular supply of paper. If you prefer to have students keep journals in composition notebooks or laboratory notebooks, have them trim the journal cover sheet to the appropriate size and paste it onto the first page of their notebooks.

4. Divide students into groups. Instruct the groups to discuss the following:
   - Ideas we have about *Earth in Space: Astronomy*
   - Questions we have about *Earth in Space: Astronomy*

   One member of the group should record his/her group's ideas and questions on a sheet of poster board or poster paper.

5. Discuss student responses by having each group, in turn, report on its ideas. As groups are responding, build up two important lists (ideas and questions) for everyone to see (on a chalk board, flip-chart paper, poster board, or an overhead transparency).

6. Direct students to add these "Ideas" and "Questions" to their journals.

7. This completes the pre-assessment phase. Distribute copies of **Investigating Earth in Space: Astronomy.**

Some sample student responses to the pre-assessment questions are provided below.

Sample Student Responses

Ideas about Astronomy
- Stars and constellations appear in the same place in the sky every night.
- The Sun rises exactly in the east and sets exactly in the west every day.
- We experience seasons because of the Earth's changing distance from the Sun (closer in the summer, farther in the winter).
- The Moon can only be seen during the night.
- The Moon does not rotate on its axis as it revolves around the Earth.
- The phases of the Moon are caused by shadows cast on its surface by other objects in the Solar System.
- The phases of the Moon are caused by the shadow of the Earth on the Moon.
- The phases of the Moon are caused by the Moon moving into the Sun's shadow.
- The shape of the Moon always appears the same.
- The Earth is the largest object in the Solar System.
- Meteors are falling stars.
- Comets and meteors are out in space and do not reach the ground.
- The surface of the Sun is without visible features.
- All the stars in a constellation are near each other.
- All the stars are the same distance from the Earth.
- The brightness of a star depends only on its distance from the Earth.
- Stars are evenly distributed throughout our Galaxy.

Ideas about Space
- The Earth is sitting on something.
- The Earth is larger than the Sun.
- The Sun disappears at night.
- The Earth is round like a pancake.
- We live on the flat middle of a sphere.
- There is a definite up and down in space.
- Seasons are caused by the Earth's distance from the Sun.
- Phases of the Moon are caused by a shadow from the Earth.

Introduction

- Different countries see different phases of the Moon on the same day.
- The amount of daylight increases each day of summer.
- Planets cannot be seen with the naked eye.
- Planets appear in the sky in the same place every night.
- Astrology is able to predict the future.
- Gravity is selective; it acts differently or not at all on some matter.
- Gravity increases with height.
- Gravity requires a medium to act through.
- Rockets in space require a constant force.
- The Sun will never burn out.
- The Sun is not a star.

Assessment Opportunity

It will be useful for you to review what your students have written before moving further into the module. This will alert you, in advance, to any specific problems they may encounter when beginning the module. Keep these lists. They also represent informal pre-assessment data, and you will be able to revisit them with your students at the end of this astronomy investigation to help track changes in understanding.

INVESTIGATING EARTH IN SPACE: ASTRONOMY

The Earth System

The Earth System is a set of systems that work together in making the world we know. Four of these important systems are:

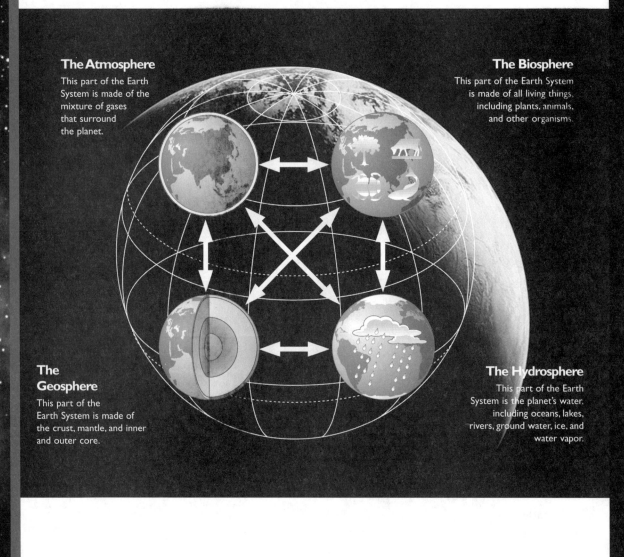

The Atmosphere
This part of the Earth System is made of the mixture of gases that surround the planet.

The Biosphere
This part of the Earth System is made of all living things, including plants, animals, and other organisms.

The Geosphere
This part of the Earth System is made of the crust, mantle, and inner and outer core.

The Hydrosphere
This part of the Earth System is the planet's water, including oceans, lakes, rivers, ground water, ice, and water vapor.

Introduction

Introducing the Earth System

Understanding the Earth System is an overall goal of the *Investigating Earth Systems* series. The fact that the Earth functions as a whole, and that all its parts operate together in meaningful ways to make the planet work as a single unit, underlies each module. Each module guides the students in considering this fundamental principle.

At the end of every investigation, students are asked to link what they have discovered with ideas about the Earth System. Questions are provided to guide their thinking, and they are asked to write their responses in their journals. They are also reminded on occasion to record the information on an *Earth System Connection* sheet. This sheet will provide a cumulative record of the connections that the students find as they work through the investigations in the module.

Not all the connections between the things they have been investigating and the Earth System will be immediately apparent to your students. They will probably need your help to understand how some of the things they have been investigating connect to the Earth System. However, by the time they complete the Investigating Earth Systems modules to the end of eighth grade, they should have a working knowledge of how they and their environment function as a system within a system, within a system...of the Earth System.

Students may, at first, struggle to find any relationship between Astronomy and the Earth System. The links are not easy to see at first. However, they will come to understand that Earth owes its very existence to its precise place in space. By looking at Earth's position in relation to the other planets of the Solar System, they will see that its unique distance from the Sun give it conditions that can support life. Planets closer to the Sun are too hot and those farther away are too cold. They will also begin to see that the Sun is the principal source of energy for planet Earth; the power source for the atmosphere, hydrosphere, geosphere, and biosphere. Students will find out that Earth and space are linked in other ways including gravitational attraction, interactions with space objects such as meteors and asteroids, and many others.

Distribute copies of the *Earth Systems Connection* sheet (**Blackline Master Astronomy I. 1**) available at the back of this Teacher's Edition. Have the students place the sheets in their journals. Explain to the students that at the end of each investigation they will be asked to reflect on how the questions and outcomes of their investigation relate to the Earth system. Tell them that they should enter any new connections that they discover on the *Earth Systems Connection* sheet. Encourage them to also include connections that they have made on their own. That is, they should not limit their entries to just those suggested in the **Thinking about the Earth System** questions in **Review and Reflect**. Use the **Review and Reflect** time to direct students' attention to how local issues relate to the questions they have been investigating. By the end of the module, students should have as complete an *Earth Systems Connection* sheet to *Investigating Earth in Space: Astronomy* as possible.

Investigating Earth in Space: Astronomy – Introduction   **23**

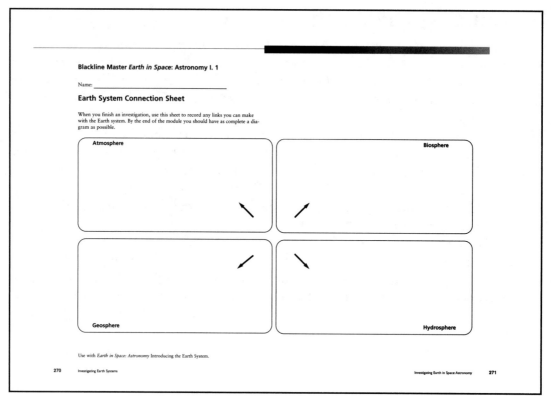

(can be photocopied onto 11x17 paper)

Introduction

Illustration By Dennis Falcon

INVESTIGATING EARTH IN SPACE: ASTRONOMY

Introducing Inquiry Processes

When geologists and other scientists investigate the world, they use a set of inquiry processes. Using these processes is very important. They ensure that the research is valid and reliable. In your investigations, you will use these same processes. In this way, you will become a scientist, doing what scientists do. Understanding inquiry processes will help you to investigate questions and solve problems in an orderly way. You will also use inquiry processes in high school, in college, and in your work.

During this module, you will learn when, and how, to use these inquiry processes. Use the chart below as a reference about the inquiry processes.

Inquiry Processes:	How scientists use these processes
Explore questions to answer by inquiry	Scientists usually form a question to investigate after first looking at what is known about a scientific idea. Sometimes they predict the most likely answer to a question. They base this prediction on what they already know or believe to be true.
Design an investigation	To make sure that the way they test ideas is fair, scientists think very carefully about the design of their investigations. They do this to make sure that the results will be valid and reliable.
Conduct an investigation	After scientists have designed an investigation, they conduct their tests. They observe what happens and record the results. Often, they repeat a test several times to ensure reliable results.
Collect and review data using tools	Scientists collect information (data) from their tests. Data can take many forms. Common kinds of data include numerical (numbers), verbal (words), and visual (images). To collect and manage data, scientists use tools such as computers, calculators, tables, charts, and graphs.
Use evidence to develop ideas	Evidence is very important for scientists. Just as in a court case, it is proven evidence that counts. Scientists look at the evidence other scientists have collected, as well as the evidence they have collected themselves.
Consider evidence for explanations	Finding strong evidence does not always provide the complete answer to a scientific question. Scientists look for likely explanations by studying patterns and relationships within the evidence.
Seek alternative explanations	Sometimes, the evidence available is not clear or can be interpreted in other ways. If this is so, scientists look for different ways of explaining the evidence. This may lead to a new idea or question to investigate.
Show evidence & reasons to others	Scientists communicate their findings to other scientists. Other scientists may then try to repeat the investigation to validate the results.
Use mathematics for science inquiry	Scientists use mathematics in their investigations. Accurate measurement, with suitable units, is very important for both collecting and analyzing data. Data often consist of numbers and calculations.

Investigating Earth Systems

Introduction

Introducing Inquiry Processes

Inquiry is at the heart of *Investigating Earth Systems*. That is why each module title begins with the word "Investigating." In the National Science Education Standards, inquiry is the first content standard. (See **Science as Inquiry** on page xii of this Teacher's Edition.)

Inquiry depends upon active student participation. It is very important to remind students of the steps in the **Inquiry Process** as they perform them. Icons that correspond to the nine major components of inquiry appear in the margins of this Teacher's Edition. They point out opportunities to teach and assess inquiry understandings and abilities. Stress the importance of inquiry processes as they occur in your investigations. Provoke students to think about *why* these processes are important. Collecting good data, using evidence, considering alternative explanations, showing evidence to others, and using mathematics are all essential to *IES*. Use examples to demonstrate these processes whenever possible.

At the end of every investigation, students are asked to reflect upon their thinking about scientific inquiry. Refer students to the list of inquiry processes as they answer these questions.

Teaching Tip

If the reading level of the descriptions of inquiry processes is too advanced for some students, you could provide them with illustrations or examples of each of the processes. You may wish to provide students with a copy of the inquiry processes to include in their journals (**Blackline Master Earth in Space: Astronomy I. 2**).

Introducing Earth in Space: Astronomy

Have you ever looked at the stars and wondered how far away they were?

Have you noticed changes in the way the Moon looks?

Did you ever wonder what it would be like to travel in space?

Have you ever looked through a telescope?

Introduction

Introducing Astronomy

This is an introduction to the module for your students. It is designed to set their investigations into a meaningful context.

Students will have a variety of experiences relating to astronomy. This is an opportunity for them to offer some of their own experiences, in a general discussion, using these pictures and questions as prompts. Some students may be able to cite additional experiences to those asked for here. Encourage a wide-ranging discussion based upon what students are able to offer from their experiences.

Since your students have just spent time in the pre-assessment thinking about and discussing what they already know about astronomy, it is probably not necessary to have them complete another journal entry. They should be eager to get to work on their investigations.

You may want to do a quick summary of the main points that emerge from the discussion. You could do this on a chalkboard, easel pad, or an overhead transparency.

About the Photos

The upper left photograph is a random view of stars in the night sky. Use this to elicit your students' visual memory of star gazing by asking them how this picture is similar and also different from what they have seen and how they might be able to tell which are closer or further away. Keep in mind that it is difficult to see stars at night from within cities or other places where there is a lot of "light pollution" (artificial lighting in street lamps or buildings).

On the upper right is a picture of a full-harvest Moon. Have students consider how the Moon has looked different at various times, especially at different times of the month.

Bottom left is a photograph of a NASA's Space Shuttle launch. Ask your students if they would like to take a trip into space, and what they think they might have to do to survive there. Do any of them think that it's likely that they may, one day, travel in space?

The final photograph (bottom right) is a professional astronomical telescope. Relatively inexpensive telescopes, for looking at the night sky, are widely available in department stores and specialist science stores. Perhaps some of your students have seen them there, or even own one. If so, suggest that they are later brought into class for all to see.

INVESTIGATING EARTH IN SPACE: ASTRONOMY

Why is Astronomy Important?

Astronomy is the study of the Moon, stars, and other objects in space. It is important because it helps you understand how the Earth system works within the Solar System and beyond. It is an amazing science because it explores space from its tiny specks of dust to its huge clusters of stars. From the beginning, humans have been fascinated by the objects in the night sky. At first, it seemed like there was just the Moon and stars, but early astronomers began to notice other objects, too. They noticed comets and meteors (called "falling stars" by early observers) moving across the skies. They also discovered that the planets were brighter and moved differently from the stars. One of the most important advances in the history of astronomy was the invention of the telescope in the 1600s. A telescope could magnify the objects in the sky. The more complex telescopes became, the more discoveries astronomers were able to make. The planet Mars looked red. Saturn had rings around it, but why? The telescope created even more questions than it answered, but astronomers kept seeking the answers. Astronomy is important because it helps you understand the universe within which Earth exists.

What Will You Investigate?

To help you understand more about the science of astronomy, you will be conducting research. Your research will include hands-on investigations. You will also make and study models. Here are some of the things you will investigate:

- what characteristics the planet Earth has
- where Earth is in our Solar System
- what other planets and objects are in the Solar System
- what the similarities and differences are between the planets
- the role of the Sun in the Solar System
- how the different parts of the Solar System relate to one another
- the objects outside our Solar System
- the life cycle of stars
- how people have studied the universe over time
- the latest theories about how our universe formed.

You will need to practice your problem-solving skills and become an accurate observer and recorder. You will also need to be creative in researching information about the universe.

In the last **Investigation** you will have a chance to apply all that you learned about Earth and space. You will create an exciting piece of communication on an astronomy topic you have learned about in this module.

Introduction

Why is Astronomy Important?

Make sure students read this section carefully. This introduction gives information from which students can conduct their own investigations throughout the module. You may want to start by having your students work in groups. Have them read this section carefully and then discuss its contents. During their discussions they can write down a series of questions about astronomy and the study of Earth and space.

Each group can present its list of questions to the rest of the class. You can then have a discussion with your students about the questions they have raised and perhaps produce a master list of questions to be posted. As the investigations proceed, and answers to these questions become known, you can add answers for all to see.

What Will You Investigate?

It is important that students get a sense of where they are headed in the module. Students need your help in making sense of the series of investigations and connecting what may seem like unrelated activities into a cohesive network of ideas. Reviewing this section of the introduction is the first step toward constructing a conceptual framework of the "Big Picture" as it is explored in *Investigating Astronomy*, including change over time and the dynamic nature of Earth and its surrounding universe. Be careful with terminology. Help them understand the difference between key words, should they arise, like: our Solar System (Earth, Sun, and the other natural objects that orbit the Sun); galaxy (a vast collection of stars and their planets held together by gravitational attraction), and universe (the whole space-time continuum in which we exist, together with all the energy and matter within it).

This would be a good time to review with students the titles of the activities in the **Table of Contents**. Ask students to explain how the titles of the activities relate to descriptions in "What Will You Investigate?" Discussing the final investigations will help students to understand the overall goal of the module. In the final investigation, students will use all they have learned to design or modify a type of technology to answer a scientific question about space.

Introducing Assessment

Students need to know, ahead of time, what expectations you have of them in terms of assessment. They need to know what they are required to do to demonstrate the highest level of study and learning. This is a good time to introduce students to assessment rubrics so that they can see how their work will be evaluated. Sample rubrics are included at the back of this Teacher's Edition.

Teacher Commentary

INVESTIGATION 1: THERE'S NO PLACE LIKE HOME

Background Information

The students begin their investigations by examining the Earth as it might be viewed from space and the various systems that occur on the Earth's surface.

There are a series of plates that move across the surface of the Earth. The plates consist of an outer layer of the Earth, the lithosphere, which is cool enough to behave as a more or less rigid shell. Occasionally, the hot asthenosphere of the Earth finds a weak place in the lithosphere to rise buoyantly as a plume, or hot spot. Only lithosphere has the strength and the brittle behavior to fracture in an earthquake. In cross section, the Earth releases its internal heat by convecting, or boiling much like a pot of pudding on the stove. Hot asthenosphere mantle rises to the surface and spreads laterally, transporting oceans and continents as on a slow conveyor belt. The speed of this motion is a few centimeters per year, about as fast as your fingernails grow. The new lithosphere, created at the ocean spreading centers, cools as it ages and eventually becomes dense enough to sink back into the mantle. The subducted crust releases water to form volcanic island chains above, and after a few hundred million years will be heated and recycled back to the spreading centers.

Water in the atmosphere moves through a cycle called the water cycle, or hydrologic cycle. Water evaporates off a body of water, such as an ocean, lake or pond. This water vapor gathers in the atmosphere until it condenses into clouds. The clouds become saturated with water and then the water falls to the ground as precipitation. This precipitation can be in the form of rain, snow, sleet, or hail depending on the temperature. When it reaches the ground some precipitation runs off into streams or lakes. That occurs if it encounters an impermeable layer which restricts it from sinking into the ground. If the precipitation does sink into the ground, it becomes part of the groundwater, a system of underground water which moves through paths within the soil and rock. The water on the surface or in the ground moves downhill until it reaches the ocean, where the system can start over again.

More Information…on the Web
Visit the *Investigating Earth Systems* web site www.agiweb.org/ies/ for links to web sites that will help you deepen your understanding of content and prepare you to teach this investigation.

Investigation Overview

Students make drawings of what they think the Earth looks like from space as well as what they think a cross-section of the Earth would look like. They then share these drawings with the rest of their classmates to come to a consensus of everyone's thoughts on the nature of the Earth. After this, students compare their drawings with a globe and diagrams of the inside of the Earth. They use this information to make a "Planet Card" of the Earth. Finally, students conduct research to answer questions they have about the Earth. They present their findings to the rest of the class. **Digging Deeper** reviews the four different systems on Earth: the atmosphere, biosphere, hydrosphere, and geosphere.

Goals and Objectives

As a result of **Investigation 1**, students will understand what the Earth looks like from space and what the Earth's interior is like.

Science Content Objectives

Students will collect evidence that shows:

1. How the Earth looks from space.
2. What the interior of the Earth is like.

Inquiry Process Skills

Students will:

1. Record observations.
2. Make a model of what they cannot see.
3. Revise models based on additional observations.
4. Communicate observations and findings to others.

Connections to Standards and Benchmarks

In **Investigation 1**, students will use their own sketches to model and make inferences about the Earth's surface and interior. Students will revise their ideas as they gather new evidence. These observations will start them on the path to understanding the National Science Education Standards and AAAS Benchmarks below.

NSES Links

- Evidence consists of observations and data on which to base scientific explanations.

- Models are tentative schemes or structures that correspond to real objects, events, or classes of events, and that have explanatory power. Models help scientists and engineers understand how things work. Models take many forms, including physical objects, plans, mental constructs, mathematical equations, and computer simulations.

Teacher Commentary

- The solid Earth is layered with lithosphere: hot, convecting mantle; and dense metallic core.

AAAS Links
- Models are often used to think about processes that happen too slowly, too quickly, or on too small a scale to be observed directly, or that are too vast to be changed deliberately, or that are potentially dangerous.

- The Earth is mostly rock. Three-fourths of its surface is covered by a relatively thin layer of water (some of it frozen), and the entire planet is surrounded by a relatively thin blanket of air.

Preparation and Materials Needed

Preparation
In **Investigation 1**, your students will create their own maps of what the Earth looks like from space and what the interior of the Earth looks like. You will need to provide students with poster board or paper as well as colored pencils and markers to complete this task.
Students will then display their posters around the room in a gallery. Make sure that there is enough blank wall space for this to happen or remove material from the walls that could become damaged during the setup of the gallery.
For the second part of the investigation, you will need to have at least one globe or relief map of Earth available for the students. Collect diagrams of the interior of the Earth as well. Higher level textbooks are a good source of these diagrams. Allow students to examine these materials to determine the accuracy of their own maps and drawings.
This investigation requires three 40-minute class periods to complete, depending upon how you structure it. **Day One:** Have the students address the **Key Question** and review the kinds of ideas they have and record them in their journals. They can then review and complete item 1 of the investigation. **Day Two:** Have students complete items 2, 3, and 4 of the investigation. **Day Three:** Have students complete items 5, 6, 7, 8 and 9, and then **Review and Reflect** upon the whole investigation.

Materials
- colored pencils or markers and blank paper
- diagram of the Earth, showing different layers
- relief map or globe of the Earth
- 5" x 8" index cards
- computer with Internet access, if possible

Investigating Earth in Space: Astronomy

Investigation I: There's No Place Like Home

Investigation I:
There's No Place Like Home

Key Question
Before you begin, first think about this key question.

Explore Questions

What characteristics does the planet Earth have and how do scientists know this?

A look at Earth from space.

Think about what you already know about the Earth. Write down two things that you feel confident you know about the Earth as a planet. You may want to draw a picture as well. Then, write down how you think scientists have discovered these things about the Earth. What tools or methods do you think they have used?

Materials Needed

For this investigation your group will need:

- colored pencils or markers and blank paper
- diagram of the Earth, showing the different layers
- relief map or globe of the Earth
- 5" x 8" index cards
- computer with Internet access, if possible

When you finish, share your ideas with other students in your group. Make a group list of what you know and what questions you would like to investigate about the Earth and how scientists study the Earth. Keep this list for later in your investigation.

Investigate

1. Use the ideas and drawings about the Earth that your group members have already finished. Work together to draw two sketches.

   a) One sketch should show the Earth as if you were looking down on it from space. Try to show as many features of the planet as you can. Label any features (continents, oceans, mountains, etc.) that you know.

Investigating Earth Systems

A
1

Teacher Commentary

Key Question

Use this question as a brief introduction to elicit students' ideas about what the Earth looks like and how scientists learned about it. Emphasize thinking and sharing ideas. Make students comfortable sharing their ideas by avoiding labeling their responses as correct or incorrect.

Write the **Key Question** on the board or on an overhead transparency. Have the students record the question and their answers in their journals. Tell students to think about and answer the question individually. Ask them to write as much as they know and to provide as much detail as possible in their responses. Emphasize thinking and sharing of ideas. Avoid seeking closure (i.e., the right answer). Closure will come through inquiry, reading relevant information, classroom discussions, and having students reflect upon what they have learned at the end of the investigation. Try to make students comfortable with sharing ideas by not commenting on the correctness of their responses.

Student Conceptions about the Characteristics of the Earth

To think about and answer this **Key Question**, your students will be drawing upon their experiences. Do not be surprised if the range of ideas is wide. Some students may have a good grasp of what the Earth is like, but others may have a more dramatic view based on science fiction movies and television. Still others may not have had an opportunity to consider the planet Earth in any organized way. It is very important that you get a sense of what all your students already know, and what they do not, before moving ahead with the investigation.

Journal Entry-Evaluation Sheet

A photocopy master of this tool is included in the Appendix. This sheet provides general guidelines for assessing student journals. Adapt this sheet so that it is appropriate for your classroom. Give a copy to students early in the module so that they can see what they are expected to include in their journals and how their entries will be assessed. You can use this tool in two ways: (a) to provide a record of student progress for you and (b) to give feedback to each student by giving them a copy of your assessment.

Answers for the Teacher Only

The Earth is composed of four interacting systems: geosphere; atmosphere; hydrosphere; and biosphere. The Earth has a layered structure consisting of: an inner core, an outer core, the mantle, and the crust. The inner and outer cores are mostly metallic. The inner core, while very hot, is mostly solid. The slightly less dense outer core is more fluid. The mantle is hot and somewhat fluid and the solid crust rests upon it.

Humans have been investigating Earth from the earliest times, observing its nature and events and searching for explanations.

INVESTIGATING EARTH IN SPACE: ASTRONOMY

Inquiry

Making Diagrams

Sometimes the best way to show the results of a scientific investigation is by drawing a diagram. Complicated concepts can often be illustrated more easily than they can be explained in words. The diagram should be labeled.

b) The second sketch should show what you and your other group members think that the inside of the Earth looks like. Imagine that you could cut the Earth in half. Do your best to show what it might look like beneath the surface. Label your sketch with descriptions of the different parts you have drawn. If you know the names of any of the Earth's layers, label these, but don't worry if you don't know this yet.

2. Label your drawings with your group's initials. Then post them up in a gallery around the room. Taking your journals, go on a "tour" around the gallery.

   a) Write down ideas that other groups have about the outside and inside of the planet.

3. When you finish the tour, talk about these ideas in your group. What ideas did you get from the other groups?

4. Now, look at the world map or globe your teacher will provide. Also look at the diagram of the inside of the Earth in the **Digging Deeper** section.

   a) How are your drawings (and those of your classmates) similar to or different from these maps and diagrams?

   b) What surprised you about the maps and diagrams?

   c) Use this new information to make your drawings as accurate as possible.

5. Next make a "Planet Card" (like a baseball card) about the Earth. Draw a picture of the Earth on one side of a 5" x 8" card and put important information that you have learned about the Earth on the back. You might include information on the following topics:

Collect & Review

Teacher Commentary

Investigate
Teaching Suggestions and Sample Answers

Assessment Tools
Key Question-Evaluation Sheet
Use this evaluation sheet to help students understand your expectations for the starting activity. The **Key Question-Evaluation Sheet** emphasizes that you want to see evidence of prior knowledge and that students should communicate their thinking clearly. You will not be likely to have time to use this assessment every time students complete a starting activity; yet, to ensure that students value committing their initial conceptions to paper and taking the activity seriously, you should always remind them of the criteria. When time permits, use this evaluation sheet as a check on the quality of their work. As with any assessment tool used in *IES*, the assessment instrument should be provided to students and discussed *before* they complete a task. This ensures that they have a clear understanding of your expectations for their work.

Assessment Tools
Journal Entry-Evaluation Sheet
Use this sheet as a general guideline for assessing student journals, adapting it to your classroom if necessary. You should give the **Journal Entry-Evaluation** Sheet to students early in the module, discuss it with them, and use it to provide clear and prompt feedback.

Investigation 1: There's No Place Like Home

- What the surface of the Earth is like.
- What the interior of the Earth is like.

As you continue reading this chapter and learn more about the Earth, you can add to the Planet Card. You can also make other Planet Cards as you learn about the other planets in the Solar System. These cards can be used as a review tool later.

6. At the beginning of this **Investigation**, you also wrote down your thoughts about how scientists found out about the characteristics of the Earth. Share your thoughts with your group. Then, as a group, choose one question that you wish to investigate further. Possible topics may include, but are not limited to, the following:

   - How did scientists discover the composition of Earth's atmosphere?
   - How do scientists know what the inside of Earth is like?
   - How deep are the oceans and how can you be sure about this?
   - How high are the highest mountain ranges on Earth?
   - What is the deepest spot in the ocean?
   - Where is the highest mountain?

7. Your class will go to the library or computer center in your school to research these topics. Be sure to ask for help from your teacher if you are having difficulty finding the information you are looking for.

8. When you finish your research, think of a way that you can present your information so that your classmates find it interesting and informative. You might want to use a PowerPoint™ presentation, an overhead transparency, a poster, photographs, or even some tools.

   a) Prepare your group's presentation.

9. Make your presentation, being sure to answer any questions from your classmates.

   a) As other people are presenting, refer to your original ideas about the Earth. Add information as you discover it. Save all this information for later investigations.

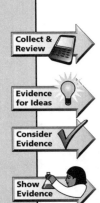

Inquiry
Scientific Questions

Science inquiry starts with a question. Scientists take what they already know about a topic, then form a question to investigate further. The question and its investigation are designed to expand their understanding of the topic. You are doing the same in this investigation.

Teacher Commentary

Journal Entry-Checklist
Use this checklist for quickly checking the quality and completeness of journal entries. You can assign a value to each criterion, or assign a "+" or "-" for each category, which you can translate into points later.

1. Students are likely to have more difficulty with the second sketch than the first one. While they may have seen images of the Earth from space, they are less likely to have seen diagrams of its interior. Some may make a connection between volcanic activity and that part of the Earth that lies beneath its surface. Movies and television may have given some students a sense of what lies below the surface, but not necessarily with scientific accuracy. Move around the room, observing what your students appear to know.
2. Using this gallery approach works well. Students can quickly see and understand what other groups have done and compare those efforts with their own.
3. It is important that the maps, globes, and diagrams you show them are large enough for students to compare with their own sketches. If necessary, they can re-draw their sketches on tracing paper, and then overlay them to see differences.
4. Students should be encouraged to use a variety of classroom resources to find out the information they need for the Planet Card. These can include books, encyclopedias, posters, and electronic media, such as the Internet.
5. This is an important activity because the Planet Card for Earth will later become the template for making planet cards for other planets. If the Internet is available, there are many web sites that show free downloadable pictures of the Earth and other planets. One option might be to collect these, print out color versions and have students glue them to the back of the index cards.
6. To avoid all groups choosing the same question to investigate, you may want to assign one question per group. It is also very important that additional questions that students themselves may have are put on the list are investigated.

Assessment Point
The list of questions students generate here will give you a good sense of their understanding and interests. Be sure to note these carefully, and perhaps refer to them at the end of the investigation.

7. It is important to allocate enough time for students to prepare and give their presentations. They will be giving presentations to their peers quite often throughout this module, so spending time in learning how to do this well is a good investment. Apart from being a very useful and transferable skill, creating an informative and interesting presentation is also a great motivator. For this reason, encourage creativity and the use of presentational tools including electronic media and software such as PowerPoint™. You should make it clear to students that you expect well thought-out, informative and complete presentations.

Investigating Earth in Space: Astronomy

INVESTIGATING EARTH IN SPACE: ASTRONOMY

Digging Deeper

EARTH, A CONSTANTLY CHANGING PLANET

Earth Systems

As You Read...
Think about:
1. What are the layers of the Earth?
2. What are the Earth's systems?

The Earth has many features and parts that work together in important ways. One way to study the Earth is to look at its different parts and understand how they are connected. With this information, you can begin to understand how the planet works and how it is always changing. Parts of the Earth that work together are known as systems. Planet Earth has four main systems: the atmosphere, the biosphere, the hydrosphere, and the geosphere.

Atmosphere

Earth as viewed from space. Which parts of the Earth systems can you see in this photo?

The picture shows Earth as viewed from space. Notice the clouds that surround the planet. They are part of an envelope of gases, called the atmosphere, that surround the Earth. When you look up into the sky from the Earth's surface, you are looking into the Earth's *atmosphere*. The gases in the atmosphere play an important role in all the Earth systems. For example, 21% of the Earth's atmosphere is oxygen. Many organisms (living things) need the oxygen in the air to live. Another important gas in the atmosphere, carbon dioxide, is used by plants to make food. Ozone is a naturally occurring gas found in a layer of the atmosphere called the stratosphere. At this level, ozone protects life on Earth from harmful energy given off by the Sun. Finally, the swirling cloud layer that you see in the photograph is condensed water vapor. This water vapor plays an important role in Earth's weather systems.

Teacher Commentary

Digging Deeper

At this stage of the activity, students read a brief passage about the Earth and its dynamic processes. This is an ideal time for you to provide students with further information about planet Earth by showing photographs, videos, books, and associated web sites. You may wish to assign the **As You Read** questions as homework.

As You Read...
Think about:
1. The layers of the Earth, from the outside to the inside are: the crust, mantle, outer core, inner core.

2. The Earth has four systems: geopshere, hydrosphere, atmosphere, and biosphere.

You can also get up-to-date information by visiting the *IES* web site www.agiweb.org/ies and its links to other useful sites.

Assessment Opportunity

You may wish to devise questions from this **Digging Deeper** section to use as quizzes. You could use multiple choice or true/false formats. This will provide assessment information about students' content understanding and can serve as a motivational tool to ensure that they complete and understand the reading assignment.

Investigation 1: There's No Place Like Home

Biosphere

The Earth supports millions of different types of living organisms that make up part of the Earth's *biosphere*. Organisms survive in many places, from high atop mountains to the extreme environments of the deep ocean floor. Some live on land surfaces, while others live below thousands of meters of glacial ice. Many organisms that once lived on the Earth no longer exist. They could not adjust when conditions such as climate and food supplies changed drastically. Fossils in ancient rocks are evidence that these organisms once did live on the Earth.

Hydrosphere

Water covers nearly 71% of the Earth's surface. The part of the Earth that contains water is known as the *hydrosphere*. Most of the water on the Earth's surface is in the oceans. Oceans are found in basins that are huge depressions in the Earth's surface. Water can be found on the Earth's land surface as streams, rivers, ponds, and lakes. Water exists underground in soil and rocks. Water in the form of vapor (gas) is an important part of the Earth's atmosphere. Water is also in the cells of every living thing on the Earth.

This image shows a massive phytoplankton bloom off the coast of Tasmania. Phytoplankton are part of the Earth's biosphere.

Geosphere

The Earth is made of layers of rock, which together make up the *geosphere*. A relatively thin layer of solid rock called the crust covers the Earth's surface. The crust has a wide variety of shapes. In some places it takes the shape of hills, mountains, slopes, or canyons. In other places it takes the shape of flatlands, shorelines, or even meteorite craters. The shape of the land is always changing. One reason for these changes is that the

INVESTIGATING EARTH IN SPACE: ASTRONOMY

The shape of Earth's crust is always changing. This image shows the mountains and valleys of the Himalaya Mountains.

Earth's massive continents are constantly being moved by processes deep within the Earth. Many of these processes occur in the part of the geosphere that lies beneath the crust. This part is called the mantle. The rocks in the mantle are continuously being squeezed, deformed, and moved in different directions. Sometimes the rocks of the crust move upward to form mountain ranges. Mountain ranges can even be found beneath the oceans and are called mid-ocean ridges.

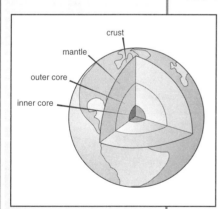

Processes deep within the Earth's interior change the shape of the landscape.

Deep beneath the Earth's mantle are two other layers: the outer core and the inner core. Both the inner and outer cores of the Earth are made mostly of iron. The inner core is solid and the outer core is liquid. These four layers – the crust, mantle, outer core, and inner core – make up Earth's geosphere.

Examining the Earth's systems provides an explanation for many of the most important features of how planet Earth works. Scientists use this knowledge when they study other planets. They compare the Earth's systems to the systems of other planets to understand how those planets work. The more they know about the Earth, the more they can learn about the other planets in the Solar System.

Investigation 1: There's No Place Like Home

Review and Reflect

Review

1. What questions about the inside and outside of the Earth were you able to answer through this investigation?
2. What questions about how scientists know about the Earth were you able to answer?

Reflect

3. What ideas about the inside and outside of the Earth surprised you the most?
4. What ideas about how scientists learn about the Earth were most interesting to you? Why were they interesting?

Thinking about the Earth System

5. What are the four main systems of Earth?
6. Describe two ways that the atmosphere and biosphere are connected. Remember to write any connections you find on the *Earth System Connection* sheet.
7. How do Earth's systems help scientists study other planets?

Thinking about Scientific Inquiry

8. In which parts of the investigation did you:
    a) Ask your own questions?
    b) Record your own ideas?
    c) Revise your ideas?
    d) Use your imagination?
    e) Share ideas with others?
    f) Find information from different sources?
    g) Pull your information together to make a presentation?

Teacher Commentary

Review and Reflect

Review
Allow your students ample time to review the investigation. Help them make all the connections based upon their data.

1. Answers will vary.
2. Answers will vary.

Reflect
Give students time to consider what they have done and the things that both surprised and interested them. It is important that they do this in a thoughtful way, looking beyond the immediate and trying to see a bigger picture of their investigations and the ways in which they learn.

3. Answers will vary.
4. Answers will vary.

Thinking about the Earth System

5. Atmosphere, hydrosphere, biosphere, and geosphere (in any order)

6.
   - Most living things in the biosphere depend on oxygen (O_2) in the atmosphere for survival.
   - Plants in the biosphere need carbon dioxide (CO_2) from the atmosphere to make food through photosynthesis.
   - During respiration, organisms take in O_2 and give off CO_2 gas into the atmosphere.
   - The atmosphere can be polluted by some human activities such as the release of auto exhaust emissions, the burning of fossil fuels, forest fires and methane from farm animals.
   - Changes in the atmosphere can affect living things.

7.
   - As far as we know, Earth is the only planet that is capable of supporting life, so understanding its complex systems provides a model that can be used for making scientific comparisons with other planets.
   - Earth's known changes over time provide a way of understanding change over time on other planets.

Thinking about Scientific Inquiry

8. Students are asked to reflect on how they used inquiry processes at the end of each investigation. This reflection helps them to build their understanding of "inquiry." Science as inquiry is a theme that runs through all investigations. Students will need many investigative experiences to grasp the many processes and skills used in scientific investigations. In this investigation, help them to understand that they:
   - started with a question to investigate
   - recorded observations and results
   - reached sensible conclusions
   - shared findings with others.

Teacher Review

Use this section to reflect on and review the investigation. Keep in mind that your notes here are likely to be especially helpful when you teach this investigation again. Questions listed here are examples only.

Student Achievement

What evidence do you have that all students have met the science content objectives?

Are there any students who need more help in reaching these objectives? If so, how can you provide this?

What evidence do you have that all students have demonstrated their understanding of the inquiry processes?

Which of these inquiry objectives do your students need to improve upon in future investigations?

What evidence do the journal entries contain about what your students learned from this investigation?

Planning

How well did this investigation fit into your class time?

What changes can you make to improve your planning next time?

Guiding and Facilitating Learning

How well did you focus and support inquiry while interacting with students?

What changes can you make to improve classroom management for the next investigation or the next time you teach this investigation?

Teacher Review

How successful were you in encouraging all students to participate fully in science learning? _____

_____

How did you encourage and model the skills values, and attitudes of scientific inquiry? _____

_____

How did you nurture collaboration among students? _____

_____

Materials and Resources

What challenges did you encounter obtaining or using materials and/or resources needed for the activity? _____

_____

What changes can you make to better obtain and better manage materials and resources next time? _____

_____

Student Evaluation

Describe how you evaluated student progress. What worked well? What needs to be improved? _____

_____

How will you adapt your evaluation methods for next time? _____

_____

Describe how you guided students in self-assessment. _____

_____

Self Evaluation

How would you rate your teaching of this investigation? _____

_____

What advice would you give to a colleague who is planning to teach this investigation? _____

_____

Investigating Earth in Space: Astronomy – Investigation 1

NOTES

Teacher Commentary

INVESTIGATION 2: THE EARTH'S MOON

Background Information

To the observer on the Earth, using only their naked eyes, relatively bright highlands and darker plains can be seen on the surface of the Moon. By the middle of the 17th century, Galileo and other early astronomers used telescopes to study the surface of the Moon, noting an almost endless overlapping of craters. Current knowledge of the Moon is greater than for any other object in the Solar System, except Earth.

On July 20, 1969, Neil Armstrong became the first man to step onto the surface of the Moon. He was followed by Edwin Aldrin, both of the Apollo 11 mission. These astrononauts experienced the effects of having no atmosphere. They had to use radio communications because sound waves cannot be heard without air. The sky around the Moon is always black because diffraction of light requires an atmosphere. The astronauts also experienced gravitational differences. The Moon's gravity is one-sixth that of the Earth's. The Moon is 384,403 kilometers from the Earth. Its diameter is 3476 kilometers. Both the rotation of the Moon and its revolution around Earth takes 27 days, 7 hours, and 43 minutes. This synchronous rotation is caused by an unsymmetrical distribution of mass in the Moon. This means that one side of the Moon is turned toward the Earth at all times. Four seismic stations were installed during the Apollo project to collect seismic data about the interior of the Moon. The results have shown the Moon to have a crust 60 kilometers thick at the center of the side closest to the Earth. The seismic determinations of a crust and mantle on the Moon indicate a layered planet with differentiation by igneous processes. There is no evidence for an iron-rich core unless it was a small one. Seismic information has influenced theories about the formation and evolution of the Moon.

The Moon was heavily bombarded early in its history, which caused many of the original rocks of the ancient crust to be thoroughly mixed, melted, buried, or obliterated. Meteoritic impacts brought a variety of rocks to the Moon from distance sources. Because the Moon has neither an atmosphere nor any water, the components in the soils do not weather chemically as they would on Earth. Rocks more than 4 billion years old still exist there, yielding information about the early history of the Solar System that is unavailable on Earth. Geological activity on the Moon consists of occasional large impacts and the continued formation of the regolith. It is thus considered geologically dead. With such an active early history of bombardment and a relatively abrupt end of heavy impact activity, the Moon is considered fossilized in time. The dark, relatively lightly cratered maria cover about 16% of the lunar surface and are concentrated on the side of the Moon closest to the Earth. Mare rocks are basalt and most date from 3.8 to 3.1 billion years. Some fragments in highland breccias date to 4.3 billion years and high resolution photographs suggest some mare flows actually embay young craters and may thus be as young as 1 billion years. The maria average only a few hundred meters in thickness but are so massive they frequently deformed the crust underneath them which created fault-like depressions and raised ridges. The relatively bright, heavily cratered areas are the highlands. The craters and basins in the highlands are formed by meteorite impact and are therefore older than the maria, having accumulated more craters.

More Information…on the Web

Visit the *Investigating Earth Systems* web site www.agiweb.org/ies/ for links to web sites that will help you deepen your understanding of content and prepare you to teach this investigation.

Investigation Overview

Students will visit three Moon stations to learn more about the features found on the Moon and the appearance of the Moon from the Earth. They will examine maps and create models at these stations to enhance their understanding of the Earth's nearest neighbor. **Digging Deeper** reviews the lunar cycle with diagrams and photographs of the Moon in its phases.

Goals and Objectives

As a result of **Investigation 2**, students will gain a better understanding of the features on the Moon, of what we are seeing when we see the Moon from the Earth, and the phases that the Moon travels through as it orbits the Earth.

Science Content Objectives

Students will collect evidence that:

1. Describes the surface of the Moon.
2. Predicts the phases of the Moon in the lunar cycle.
3. Explains the possible origins of the craters on the surface of the Moon.

Inquiry Process Skills

Students will:

1. Investigate using a variety of information and data resources.
2. Use models to reveal ideas and information.
3. Make observations and record results.
4. Communicate observations and findings to others.

Connections to Standards and Benchmarks

In **Investigation 2**, students investigate evidence to develop an understanding of the features of the Moon's surface and lunar phases. The evidence they review and collect will set the stage for understanding about the general structure and properties of the Earth, its Moon, the Sun, and the rest of the Solar System. These observations and investigations contribute to developing the National Science Education Standards and AAAS Benchmarks below:

NSES Links

- Evidence consists of observations and data on which to base scientific explanations.
- The motion of an object can be described by its position, direction of motion, and speed.

Teacher Commentary

- The Earth is the third planet from the Sun in a system that includes the Moon, the Sun, eight other planets and their moons, and smaller objects, such as asteroids and comets. The Sun, an average star, is the central and largest body in the Solar System.
- Most objects in the Solar System are in regular and predictable motion. Those motions explain such phenomena as the day, the year, phases of the Moon, and eclipses.

AAAS Links

- The Moon's orbit around the Earth once in about 28 days changes what part of the Moon is lighted by the Sun and how much of that part can be seen from the Earth—the phases of the Moon.

Preparation and Materials Needed

Preparation
The first part of this investigation requires the students to visit three different Moon stations. Depending upon the number of students in your class, you may find it useful to set up two different stations for each of the Moon Stations. This would allow six groups of student to be working on an activity at the same time.

This investigation is designed to be student-driven group work. Students specialize in learning a particular aspect about the Earth's Moon. At the end, you will need to organize a whole-class discussion about what student groups learned from each station.

This investigation requires four 40-minute class periods to complete, depending upon how you structure it. **Day One:** Have the students address the **Key Question** and review the kinds of ideas they have and record them in their Journals. They can then review the investigation and discuss how they will go about building their stations. **Day Two:** Have students construct their Learning Stations. **Day Three:** Have students visit each other's Learning Stations and then hold a class discussion about what everyone has observed and learned. **Day Four:** Review the **Digging Deeper** section and complete the **Review and Reflect** items.

Materials Needed
Moon Station 1
- Moon map
- tape

Moon Station 2
- Moon-phase diagram
- lamps or flashlights
- tennis ball
- Ping-Pong™ ball
- pencil
- overhead projector

Moon Station 3
- deep metal or plastic container
- flour
- index card
- metric ruler
- meter stick
- small items of varying sizes
- learning station materials (poster board, markers, blank paper, colored pencils)

Teacher Commentary

NOTES

INVESTIGATING EARTH IN SPACE: ASTRONOMY

Investigation 2:

The Earth's Moon

Key Question
Before you begin, first think about this key question.

What are the features of the Moon?

Share what you know about what the Moon looks like. Keep a record of your ideas in your journal.

Share your group's ideas with the rest of the class.

The universe is so vast that it is a very difficult job to study all its parts in depth. Scientists can start with the Earth, as you did in this module. From there, one method of learning more about the universe might be to investigate Earth's nearest neighbor, the Moon. Another method might be to study the object on which we depend the most for energy, the Sun. Scientists also study the relationships among Earth, Moon, and Sun.

Investigation 2 will get you started by focusing on the Moon.

Materials Needed

For Station 1 your group will need:

- Moon map
- tape (optional)
- Learning Station materials (poster board, markers, blank paper, tape, stapler)

Investigate
Your class will work in specialist groups to study the Moon. Your group will then set up Learning Stations for other class members.

Moon Station 1: Moon's Features

1. Spread out the Moon map (or tape it up) so that you can see all the information on the map. Observe the main types of features on the Moon.

Teacher Commentary

Key Question
Begin by asking students to respond to the **Key Question**, "What are the features the Moon?" Tell students to write their ideas in their journals. After a few minutes, discuss students' ideas in a brief conversation. Emphasize thinking and sharing of ideas. Avoid seeking closure (i.e., the "right" answer). Record all of the ideas that the students share on the board or an overhead transparency. Have students record this information in their journals.

Student Conceptions about the Earth's Moon
Encourage students to describe observations they have made of the Moon. Try to draw out information from the students by asking a series of questions. For example, ask them if they have ever seen the Moon low in the sky and if they have ever seen the Moon in the daytime. This will get the students thinking about the Moon and what we may be seeing from the surface of the Earth.

Some students may think that the far side of the Moon is the same as the dark side of the Moon. In actuality, although the same side of the Moon always faces the Earth, the Moon's position with respect to the Sun is not fixed. As the Moon revolves around the Earth, sunlight shines on the near and far sides of the Moon at different times.

Answer for the Teacher Only
Features on the Moon include the maria (low, dark areas of basaltic flows), the highland (light colored areas), rilles, and craters.

> ### Assessment Tools
>
> ### Journal Entry-Evaluation Sheet
> A photocopy master of this tool is included in the Appendix.

Investigate

Teaching Suggestions and Sample Answers

Moon Station 1 – Moon's Features
1. Students should be able to recognize features such as maria, rilles, highlands, craters, rays, and lunar regolith on their Moon maps. Review the characteristics of each of these features with the students.

   Maria – extensive dark areas that represent great basins on the Moon
   Rilles – long, deep cracks within the Moon's maria regions
   Highlands – upland areas
   Craters – depressions
   Rays – bright streaks of shattered rock and dust that radiate from a number of the Moon's craters
   Lunar regolith – lunar soil

Investigation 2: The Earth's Moon

2. Try to answer all the following questions about the Moon map. You can use this information in your Learning Station.

   a) What differences do you observe between how the two sides of the Moon look on the map? How could you explain these differences?

   b) What are the Moon's maria? How might they have formed?

   c) How do you think the craters formed on the Moon?

   d) Do you see any evidence that there once might have been water on the Moon? Do you see evidence that there once might have been life on the Moon? Explain your answer.

   e) How does the Moon's surface compare to the Earth's? (Look at a relief map of the Earth, or a globe.)

3. When you have finished studying the Moon map and answering the questions, think about how you could build a Learning Station to help others in your class learn more about the Moon's features.

   a) Record your plan and show it to your teacher.

Moon Station 2: Moon Phases

1. Study the Moon-phase diagram carefully. How does the Moon appear to change, as you observe it from the Earth, over the course of a month? How could you explain this, thinking about the relationships among the Moon, the Earth, and the Sun? (You might want to look at a diagram of the Solar System.)

2. Now, look over the materials you can use to make a model of how the Moon appears to change shape. Think about these questions before you make your model:

   a) What object(s) do you have in your materials set that could represent the Sun? The Earth? The Moon?

Materials Needed

For Station 2 your group will need:

- Moon-phase diagram
- strong lamps or flashlights
- Ping-Pong® ball
- pencil
- tape
- overhead projector (optional)
- Learning Station materials (poster board, colored pencils, blank paper, tape, stapler)
- tennis ball

Phases of the Moon

Teacher Commentary

2. a) The front side of the Moon is nearly half highlands and half maria. The back side of the Moon is mostly highlands and craters with very few regions of maria.
   b) The maria regions are areas of basalt flow, indicating past volcanic activity on the Moon. In the past, scientists believed that the maria were bodies of water but we now know this is not true.
   c) The Moon's craters were formed by debris hitting the surface during the early stages of the Moon's development. The lack of erosional forces has kept them there.
   d) In the past, scientists believed that the maria were bodies of water but we know this is not true. The rilles on the surface appear to be somewhat like dry riverbeds. There is no other evidence, however, indicating that there was ever water on the Moon. Therefore, there is no evidence of life on the Moon.
   e) The surface of the Moon is barren compared to the surface of the Earth. The Moon is littered with craters while not nearly as many craters exist on the Earth.

3. Student plans will vary. Be sure that the plans are well developed, and doable in the time period and constraints the students have to work with.

Moon Station 2 – Moon Phases

Student models should use the lamp as the Sun, the tennis ball as the Moon, and the students themselves as the Earth. Students should walk through the different phases as part of their model, indicating where each phase would be. You may wish to darken the room to make the lighted side of the table tennis ball more apparent.

> **Teaching Tip**
>
> Remind students that if they were standing on the surface of the Moon, the Earth would go through a complete cycle of phases. When an observer on the Earth sees a new Moon, an observer on the dark side of the Moon would see a full Earth shining back at them.
>
> Explain that during the first crescent Moon, observers can often see a faint glow from the unlit part of the Moon. This is caused by the light from the bright Earth being reflected off the Moon.

INVESTIGATING EARTH IN SPACE: ASTRONOMY

b) Which objects would you have to move and which will you keep still?

c) How might you move those objects to show how the Earth and the Moon move?

3. You might find it useful to draw a diagram of the Sun/Earth/Moon system before building your model.

a) Draw a diagram of your model and submit it to your teacher for approval. Use the model in your Learning Station.

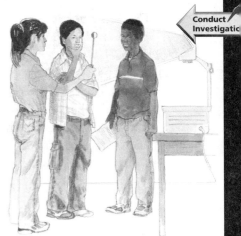

Materials Needed

For Station 3 your group will need:

- Moon map
- deep metal or plastic container (such as a rectangular cake pan or storage bin)
- flour
- index card
- metric ruler
- meter stick
- small objects of various sizes and shapes (golf ball, wooden block, rock, etc.)
- Learning Station materials (poster board, colored pencils, blank paper, tape, stapler)

Moon Station 3: Moon Craters (optional)

1. Study the Moon map carefully. Note its features, particularly the craters (depressions).

a) What are the characteristics of the craters (shape, depth, special features)?

b) How do you think they formed?

c) What shape and type of object most likely formed the craters?

2. Make a model of how craters formed. Fill the pan to a depth of 2.5 cm (1") with flour. This will represent the Moon's surface before it was cratered. Now, think about how you can use the collection of objects (ball, block, etc.) to make similar craters in the flour.

a) When you have a plan, write it down and submit it to your teacher for approval.

3. Test out your plan.

a) Be sure to record what happens and what the craters look like. You might also want to measure both the objects and the craters they form.

b) How do the craters you made in the flour compare to the Moon's craters on the Moon map? How are they different?

c) What object was the most likely shape to have made the craters? What evidence do you have for that?

Teacher Commentary

Assessment Tools

Key Question-Evaluation Sheet

Use the evaluation sheet to help students understand your expectations for the starting activity. **The Key Question-Evaluation Sheet** emphasizes that you want to see evidence of prior knowledge and that students should communicate their thinking clearly. You will not likely have time to apply this assessment every time students complete a starting activity; yet, in order to ensure that students value committing their initial conceptions to paper and taking the activity seriously, you should always remind them of the criteria. When time permits, use this evaluation sheet as a check on the quality of their work. As with any assessment tool used in *IES*, the assessment instrument should be provided to students and discussed *before* they complete a task. This ensures that they have a clear understanding of your expectations for their work.

Moon Station 3 – Moon Craters

Before students begin this activity, ask them to write in their journals the other objects in the Solar System that have craters. Ask them if they know of any craters in the United States. Explain that Mercury, the Moon, and many other moons are covered in craters. This is because they have little or no atmosphere. That means there is no direct source of mechanical or chemical weathering so the surface does not become smooth. Weathering of the Earth's surface and plate tectonics has caused the disappearance of many craters that once covered the surface.

Remind students to wear safety goggles for this activity. It should also be emphasized that the objects should be dropped, not thrown, into the flour.

Students should realize that the craters they form in the flour will be very fragile. They should take special caution not to breathe on or move the pan after dropping the object so they can make their observations.

Assessment Tools

Journal Entry-Evaluation Sheet

Use this sheet as a general guideline for assessing student journals, adapting it to your classroom if desired. You should give the **Journal Entry-Evaluation** Sheet to students early in the module, discuss it with them, and use it to provide clear and prompt feedback.

Journal Entry-Checklist

Use this checklist for quickly checking the quality and completeness of journal entries. You can assign a value to each criterion, or assign a "+" or "-" for each category, which you can translate into points later.

Investigation 2: The Earth's Moon

4. When you have finished working with your crater model, figure out how you can design and build a Learning Station to help others understand how the Moon's craters were formed.

Building and Visiting the Learning Stations

1. After your teacher has approved your plans, prepare your Learning Station. Make sure it includes the following:

   - name of your station (Example: Moon Phases)
   - objective(s) for the station (Example: This station is designed to help visitors understand how the phases of the Moon change.)
   - materials for other groups to do part of the investigation at the station (Example: maps, models, diagrams)
   - procedure for the investigation (guidelines for how to do the investigation)
   - two questions for groups to answer or two tasks to perform to check their understanding (Example: Draw and label three phases of the Moon.)
   - scientific explanation of the important concepts for the station. (Example: A new moon occurs when . . .)

   You might find it useful to have explanation handouts to give to visitors to your center.

Inquiry

Sharing Findings

An important part of science inquiry is sharing the results with others. Scientists do this whenever they think that they have discovered interesting and important information. This is called disseminating research findings. In this investigation you are sharing your findings with other groups.

2. When all the Learning Stations are set up, your teacher will work with you to organize group visits to all stations. You will need to take your journals with you to record your answers to the questions at each station. Your journal will also help organize any handouts other groups may have prepared for you to take away. When you finish each station, be sure to set it back up as you found it so that other groups have the same experience you did.

   a) Answer the questions for each station in your journal.

3. When everyone finishes all the Learning Stations, have a whole-class discussion about what you learned from each station. Check your understanding of the key ideas from each station with other groups and with your teacher. Write down other questions about the Moon that you can investigate on your own later on.

Teacher Commentary

NOTES

Investigating Earth in Space: Astronomy

INVESTIGATING EARTH IN SPACE: ASTRONOMY

As You Read...
Think about:
1. Why does the Moon look different from Earth during the lunar cycle?
2. When does the new Moon occur?
3. When does the full Moon occur?

Digging Deeper

THE LUNAR CYCLE

As you discovered in this investigation, the Moon's appearance changes during its orbit around the Earth. This series of phases is called the *lunar cycle*. You can only see the Moon when sunlight is reflected from its surface. The same side of the Moon always faces the Earth. As the Moon orbits the Earth, the angle between the Sun, the Earth, and the Moon changes. As this angle increases, you can see more of the Moon. Scientists can predict lunar cycles because the directions and speeds of the orbits of the Earth and Moon are very well understood.

Lunar phases depend on how the Earth, Moon, and Sun are positioned relative to one another. When the Moon is located between the Earth and the Sun, the side of the Moon that is illuminated is the side facing away from the Earth so you do not see the Moon.

Moon, Earth, Sun as viewed from above our Solar System

diagram not to scale

The relative positions of the Earth, Moon, and Sun determine the lunar phases.

Teacher Commentary

Digging Deeper
This section provides text and photographs to give students greater insight into the Moon. You may wish to assign the **As You Read** questions as homework to help students focus on the major ideas in the text.

As You Read...
Think about:
1. The Moon looks different because different visible parts are lit by the Sun.
2. A new Moon occurs when the Moon is directly between the Earth and the Sun. None of the visible portion of the Moon is lit.
3. A full Moon is when the Earth is between the Moon and the Sun. The entire visible side is lit.

The Lunar Cycle
This area is a complex set of scientific concepts. Initially, it may be helpful to divide this **Digging Deeper** into sections and have different groups of students specialize in them. However, it will be important that everyone has an opportunity to understand the full picture.

Be sure to allow time for a full discussion after students study the **Digging Deeper** section. Be alert for any student misunderstandings and find ways to clarify the concepts involved.

Assessment Opportunity
You may wish to rephrase selected questions from the **As You Read** section into multiple choice or true/false format to use as a quiz. Use this quiz to assess student understanding and as a motivational tool to ensure that students complete the reading assignment and comprehend the main ideas.

Investigation 2: The Earth's Moon

This phase is called the *new* Moon. During the first half of the lunar cycle, you see a little more of the illuminated Moon each time it rises. When the Earth, the Moon, and the Sun are in the same plane and nearly in a line, you see either a new Moon or a *full* Moon.

The cycle of lunar phases takes about 29.5 days to complete. After a new Moon, you gradually see more and more of the Moon until, about two weeks after the new Moon, a full moon appears in the sky. At this point, the entire visible side of the Moon is illuminated by the Sun. Over the two weeks that follow, you see less and less of the Moon until it is a new Moon again.

Teacher Commentary

NOTES

INVESTIGATING EARTH IN SPACE: ASTRONOMY

Review and Reflect

Review
1. Describe the main features of the Moon.
2. How can scientists predict lunar cycles?

Reflect
3. Explain why the lunar cycle is called a cycle.

Thinking about the Earth System
4. How can the craters on the Moon help scientists better understand the geosphere?

Thinking about Scientific Inquiry
5. How did you use modeling in this investigation?
6. Why do you think sharing findings is an important process in scientific inquiry?

Teacher Commentary

Review and Reflect

Review

Be sure to give students time to review what they have done and understood in this investigation. Ensure that all students think about and discuss the questions listed here. Be alert to any misunderstandings and, where necessary, help students to clarify their ideas.

1. The Moon has dark areas, called maria, which were formed from basalt flows. The areas that appear light from the Earth are the rough, cratered lunar highlands. Rilles are long channels found within the maria. And the rough, lunar soil is called regolith.

2. The lunar cycles can be predicted by understanding how the Moon and the Earth move with respect to each other. The Moon is rotating on its own axis at about the same speed as the Earth rotates on its axis. This means that we always see the same side of the Moon. As the Moon revolves around the Earth in a period of 28.5 days, it moves trough a series of phases, depending on where the Sun is in relation to the visible side of the Moon.

Reflect

3. The lunar cycle is a cycle because it goes through the same phases every 28.5 days.

Thinking about the Earth System

4. The Moon's craters have been caused by past collisions of space debris (meteors, asteroids, etc.) with the Moon's surface. Because these craters have not changed over geologic time, scientists can interpret them to better understand the impact of space debris that collides with Earth's surface.

Thinking about Scientific Inquiry

In this investigation, your students have been modeling experiments to clarify the relationship between the Earth and the Moon. Have your students carefully consider the list of Science Processes shown on page x of their edition and record which of them they have used in this investigation. (**A Blackline master Astronomy 1.2**, is included in the Appendix.)

5. In this investigation, modeling is used to show how the Moon moves around the Earth in its orbit. This model created a simulated month in which the different phases of the Moon were passed through.

6. Sharing findings is important to the scientific process because it allows other scientists to see what you have found. This will expand the body of knowledge about a topic, allow other people to duplicate your experiments, as well as warn others to the possible difficulties they may encounter.

Teacher Review

Use this section to reflect on and review the investigation. Keep in mind that your notes here are likely to be especially helpful when you teach this investigation again. Questions listed here are examples only.

Student Achievement

What evidence do you have that all students have met the science content objectives?

Are there any students who need more help in reaching these objectives? If so, how can you provide this?

What evidence do you have that all students have demonstrated their understanding of the inquiry processes?

Which of these inquiry objectives do your students need to improve upon in future investigations?

What evidence do the journal entries contain about what your students learned from this investigation?

Planning

How well did this investigation fit into your class time?

What changes can you make to improve your planning next time?

Guiding and Facilitating Learning

How well did you focus and support inquiry while interacting with students?

What changes can you make to improve classroom management for the next investigation or the next time you teach this investigation?

Teacher Review

How successful were you in encouraging all students to participate fully in science learning? _____

_____

How did you encourage and model the skills values, and attitudes of scientific inquiry? _____

_____

How did you nurture collaboration among students? _____

_____

Materials and Resources

What challenges did you encounter obtaining or using materials and/or resources needed for the activity? _____

_____

What changes can you make to better obtain and better manage materials and resources next time? _____

_____

Student Evaluation

Describe how you evaluated student progress. What worked well? What needs to be improved? _____

_____

How will you adapt your evaluation methods for next time? _____

_____

Describe how you guided students in self-assessment. _____

_____

Self Evaluation

How would you rate your teaching of this investigation? _____

_____

What advice would you give to a colleague who is planning to teach this investigation? _____

_____

NOTES

Teacher Commentary

INVESTIGATION 3: THE EARTH AND ITS MOON
Background Information

In one sense, all students know about gravity. They know that objects, including themselves, usually fall downwards toward the ground. They know the old saying "What goes up, must come down!" However, they may not understand that gravity is a "pulling" force; seeing it more as a "push." The idea that an object has gravity and because an object has mass, which makes it attracted to, and be attracted by, other objects can seem very confusing and is difficult to understand. Combining this with the variables of the relative "amount" of the mass and the "distance" between two objects adds to the complexity. One way to help students is for them first to understand that all matter is attracted to all other matter. Most of the time we don't see this on Earth because the objects are small masses. For example, two tennis balls lying a few inches apart are both exerting gravitational pull, but their mass is so small that they do not move toward each other and they are clinging to the surface of the tennis court because the "pull" from Earth is much stronger.

Isaac Newton's ideas about gravity were published in 1687. He recognized that all objects have gravity: some have a lot and some very little, depending upon their relative masses. The more massive an object is the more gravity is has. The closer two objects are, the stronger the gravitational pull between them. Putting these two things together, the more massive and close two objects are, the greater the gravitational pull between them.

If we think of the Earth and its gravitational pull on the Moon, and the Moon's pull on the Earth, we might expect them to collide. Your students may be confused about this, because it seems to deny the logic of gravitational pull. They need to understand that the Moon is in motion and its mass is much less than that of Earth. It is in orbit around the Earth and this force of motion acts against the gravitational pull, keeping the two objects linked together but never touching.

Many students may not have a scientifically accepted understanding of gravity. They may, for example, think that gravity is a force that "pushes" objects down toward the surface of the Earth. They may not understand that the gravity of a mass (like the Earth, Sun or Moon) attracts and exerts a "pulling" force on objects, towards its center.

More Information…on the Web
Visit the *Investigating Earth Systems* web site www.agiweb.org/ies/ for links to a variety of web sites that will help you deepen your understanding of content and prepare you to teach this investigation.

Investigation Overview

In **Investigation 3**, students learn more about the relationship between the Earth and the Moon. They complete a series of investigations which are all part of a larger GravLab, designed to look at the relationship between gravity on the Earth and gravity on the Moon. Students will compare the weight on an object on the Moon to the weight of that object on Earth, determine how far they can jump on the Moon, and learn how tides are determined by the Moon's gravitational pull on the Earth. **Digging Deeper** reviews the role of gravity in the organization of the Solar System as well as a review of tides on the Earth

Goals and Objectives

As a result of **Investigation 3**, students will have a greater understanding of the gravitational relationship between the Earth and the Moon.

Science Content Objectives

Students will collect evidence that:
1. Gravity differs on the Earth and on the Moon.
2. Distances that a person can jump differ between the Earth and the Moon, because of gravitational differences.
3. The gravitational pull of the Moon on the Earth creates the tides.

Inquiry Process Skills

Students will:
1. Create models to answer questions.
2. Collect data from the models.
3. Analyze evidence from the models.
4. Arrive at conclusions based on the evidence.

Connections to Standards and Benchmarks

In **Investigation 3**, students explore the relationship between the Earth and the Moon and then engage in activities to understand gravity. The observations and investigations they complete will help them understand the National Science Education Standards and AAAS Benchmarks below:

NSES Links

- The Earth is the third planet from the Sun in a system that includes the Moon, the Sun, eight other planets and their moons, and smaller objects, such as asteroids and comets. The Sun, an average star, is the central and largest body in the Solar System
- Most objects in the Solar System are in regular and predictable motion. Those motions explain such phenomena as the day, the year, phases of the Moon, and eclipses.
- Gravity is the force that keeps planets in orbit around the Sun and governs the rest of the motion in the Solar System.

Teacher Commentary

AAAS Links

- Nine planets of very different size, composition, and surface features move around the Sun in nearly circular orbits. Some planets have a great variety of moons and even flat rings of rock and ice particles orbiting around them. Some of these planets and moons show evidence of geologic activity. The Earth is orbited by one Moon, many artificial satellites, and debris.

Preparation and Materials Needed

Preparation

GravLab

You may want to bring in a tennis ball and toss it in the air to demonstrate the pull of gravity. Often, it helps students who are visual or kinesthetic learners to see or do something to understand a concept.

Mini-Investigation A

Aside from the text, this part of **Investigation 3** requires little advance preparation.

Mini-Investigation B

You will need to find an area to conduct this investigation that is open enough for students to jump but that does not disturb other classrooms. An isolated hallway or an outside area may work well.

If there are students in the class who are unable to jump due to physical disabilities or personal preference, assign them the role of measurer or recorder. Make sure that every member of the group has a role or a task.

If you are unable to find an appropriate location to perform this activity or you are pressed for time, it is possible to give the students a set of sample data for them to manipulate. This is not as meaningful or memorable for the students but it will allow you to get the point across.

Mini-Investigation C

You may want to bring in photographs of an area at high tide and then the same area at low tide to show students the difference.

Actual tide tables are found in many newspapers or on the Internet. Encourage students to look for actual tide tables to see how they change.

This investigation requires five 40-minure class periods to complete, depending upon how you structure it. **Day One:** Have the students address the **Key Question**, review the ideas they have and record them in their Journals. They can then work through the **Gravlabs** introduction section items 1 and 2. **Days Two and Three:** Complete Mini-Investigation A and Mini-Investigation B. **Day Four:** complete **Mini-Investigation C** and the **Sharing and Discussing Your Findings** section. **Day Five:** Have students read the **Digging Deeper** section and **Review and Reflect** on the whole investigation.

Materials

Mini-Investigation A:
- calculator

Mini-Investigation B:
- masking tape
- metric measuring tape
- calculator

Teacher Commentary

NOTES

Investigation 3: The Relationship between the Earth and Its Moon

Investigation 3:
The Relationship between the Earth and Its Moon

Key Question
Before you begin, first think about this key question.

What is the relationship between the Earth and its Moon?

Share what you know about the relationship between the Earth and the Moon with others in your group. Keep a record of your ideas in your journal.

Share your group's ideas with the rest of the class.

Investigate
GravLab

The purpose of this **Investigation** is to help you understand how the force of gravity works on the Earth. It will also get you to begin thinking about the role of gravity in the Solar System and beyond. As you go through each part of the **Investigation**, think about how you are building your understanding of gravity.

1. Think about the following question: What is gravity? Talk about this with other members of your group. Here are some questions to think about during your discussion:
    - How high can you jump off the ground? What happens to you when you get to the top of your jump?

Teacher Commentary

Key Question

Use the **Key Question** as a brief warm-up activity to elicit students' ideas about the relationship between the Earth and its Moon. Write the question on the board or on an overhead transparency. Have students record the question and their answers in their journals. Discuss students' ideas. Ask for a volunteer to record responses on the board or on an overhead transparency. Circulate among the students, encouraging them to copy the notes in an organized way.

Assessment Tool
Key Question-Evaluation Sheet
Use this evaluation to help students understand and internalize basic expectations for the warm-up activity. The **Key Question-Evaluation Sheet** emphasizes that you want to see evidence of prior knowledge and that students should communicate their thinking clearly. You will not likely have time to apply this assessment every time students complete a warm-up activity; yet, in order to ensure that students value committing their initial conceptions to paper and are taking the warm up seriously, you should always remind them of the criteria. When time permits, use this evaluation sheet as a spot check on the quality of their work.

Student Conceptions about the Earth and the Moon
Students are likely to understand the role gravity plays in their life on the Earth and that the Moon is held into orbit by the Earth's gravity. But they may not understand that the Moon also has a gravitation force that has an impact on things on the Earth.

Answer for the Teacher Only
The Earth and the Moon both have a gravitational pull which has an impact on the other. The Moon is held in orbit, largely by the gravitational pull of the Earth. The Moon also exerts a pull on the Earth, changing things like tides.

Investigate

Teaching Suggestions and Sample Answers

GravLab
These questions are designed to be a springboard for your students to discuss and learn more about gravity in relation to Earth, Moon and Sun. Ensure that they discuss them carefully and add any new questions. Their task will be to find answers to these questions that are scientifically accurate.

1. Student responses will vary. Check in with the student groups to make sure that they are on the right track with their answers. The discussion and definition that students have here is limited to their own knowledge base. It's important that they establish what they know before researching for further, more accurate

INVESTIGATING EARTH IN SPACE: ASTRONOMY

- What happens when you throw a ball into the air? How can you explain this?
- Think about any rockets you have seen on television taking astronauts into space. Why do they need so much power to get off the launch pad?
- Think about pictures of astronauts in space capsules. What is unusual about the way they move around? Why do you think this is so?

a) When you finish the discussion, write down a whole-group explanation of what you think gravity is. Do your best to agree as much as possible, but it is all right if people have different ideas at this point.

2. When you feel comfortable with your explanation, share it with the rest of the class to see what ideas other groups can add to yours. Your teacher will record the class's definition of gravity on the board or on chart paper.

a) Record the class's definition in your journal.

You will now work with your group to investigate gravity in a number of ways. When you finish, you will revisit your class definition of gravity to see how it might need to change.

Mini-Investigation A: Gravity on the Moon

1. Look over the information in the table on the next page for this **Mini-Investigation**. It shows how much common objects weigh on Earth and how much they would weigh on the Moon.

   a) Write a sentence that explains the relationship between the weight of an object on Earth and the weight of the same object on the Moon. You do not have to use an exact number, just write a description of how the weights compare.

2. When you have thought about the relationship, check with another group to see what they have discovered. Talk this over as a class with the help of your teacher. Write your answers to the following questions:

Teacher Commentary

information. In this way they have a stake in finding out how reasonable their own explanations are, and will also be more focused on their research to find out more. Resist any temptation to correct these initial ideas.

2. Here, the group is bouncing their collective ideas against those held by other members of the class. Again, let this go the way it goes, resisting any temptation to agree or disagree with the ideas or explanations that students give. This will come later.

Mini-Investigation A: Gravity on the Moon

An object's weight depends on the object's mass and the mass and radius of the large body it is near. Thus, in general, objects will have different weights on planets with different masses and radii.

1. Student responses will vary. They should indicate something similar to the following: "The weight of an object on the Earth is much greater than the weight of the same object on the Moon." Students may also recognize that the weight of something on the Earth is roughly 6 times greater than it is on the Moon.

Investigating Earth in Space: Astronomy

Investigation 3: The Relationship between the Earth and Its Moon

Table: Weight of Objects on Earth and on the Moon		
Object	Weight on Earth (kg)	Weight on the Moon (kg)
Adult male African elephant	6800	1134
Pair of adult man's tennis shoes, size 9	1.2	0.2
Gallon of paint	6.6	1.1
3-L tin of cooking oil	3.0	0.5

a) Why do you think there is this relationship? Think about the size of the Earth compared to the size of the Moon. Does that help?

b) As a group, write one sentence explaining what you think the relationship is between gravity on the Moon and gravity on the Earth.

Mini-Investigation B: You're on the Moon! Now, Jump!

In this part of the **Investigation**, you will be figuring out how far you could jump if you were on the Moon. Think about **Mini-Investigation A** and the relationship you discovered between the gravity on the Earth and on the Moon. Now, imagine how that might affect how far you could jump on the Moon.

Materials Needed

For this part of the investigation your group will need:

- masking tape
- measuring tape (metric)
- calculator

1. To get a better idea of this, first see how far you can jump from a standing position on the Earth.

Investigating Earth Systems

Teacher Commentary

2. a) Answers will vary but students may understand that the Earth is much larger than the Moon. This may mean that there is a greater pull of gravity which in turn makes the weight of something greater on the Earth.
   b) Answers will vary.

Mini-Investigation B: You're on the Moon! Now, Jump!

1. To do this, students need to make a connection between gravity and motion (in this case their own jumping). Jumping is essentially using muscle power as an opposing force to that of gravity. At first, the jump exerts enough force to overcome the opposing force of gravity, but it cannot be sustained and rapidly begins to fade until gravity, once again, takes over and pulls the jumper back toward the center of the Earth.

Investigating Earth in Space: Astronomy

INVESTIGATING EARTH IN SPACE: ASTRONOMY

a) Make a data table for your group. The table should hold the following information for each student in the group: distance jumped on Earth, estimated distance on the Moon, estimated distance on Jupiter.

2. Take turns being *Jumper*, *Measurer*, and *Recorder*. Follow these steps:

   - In a hallway or the gym, use masking tape to mark a starting line.
   - *Jumper 1* stands with toes on the starting line and jumps as far as possible.
   - *Measurer 1* holds the end of the measuring tape on the starting line.
   - *Measurer 2* pulls the tape to where the Jumper lands and puts a second piece of masking tape on the floor.
   - *Measurer 2* records the Jumper's initials and length of jump in centimeters on the masking tape at the end of the jump.
   - The *Recorder* records the initials and distance in the data table for the group. The *Recorder* should also be on hand to steady the *Jumper* so that he or she doesn't fall at the end of the jump.
   - After *Jumper 1* jumps, everyone rotates roles so that all group members get a chance to jump.

3. Remember the relationship between the weight of objects on Earth and the same objects on the Moon. To find the weight of an object on the Moon, you can multiply the weight of that object on the Earth by 0.1667.

   For example:

   The weight of an elephant on Earth is 6800 kg.

   The weight of the elephant on the Moon is
   6800 kg × 0.1667 = 1133.56 kg or about 1134 kg.

   Use this procedure to determine how far each person could jump on the Moon.

   a) Place this information in the data table.

4. Measure these distances for each *Jumper* on the floor. Mark each distance with a new piece of masking tape with initials and the new distance in centimeters. Label each of these tape pieces with MJ (Moon Jump).

Inquiry

Using Mathematics

Mathematics is often used in science. In this investigation you began by comparing the weight of an object on Earth to the weight of the object on the Moon. Then you used the mathematical relationship to calculate how far you can jump on the Moon.

A 18

Investigating Earth Systems

Teacher Commentary

2. The same principle as mentioned for **Step 1** applies to a long jump, rather than a straight-up jump. Here, however the jump has some forward motion as well as upward motion. A successful long jumper has to balance the speed of the forward motion (the run-up) with upward motion (the take-off). Doing one without the other will not produce a result. Getting the best of both in harmony will produce the longest distance possible for that person.
3. Make sure students are clear about the procedures here. Encourage them to try everything out a few times before starting the actual investigation and measurements of jumping.
4. As members of the group exchange tasks ("Jumpers", "Measurers" and "Recorders") they should perform their tasks in exactly the same way as others do. It's important that students see the need to control variables as much as reasonably possible.

Assessment Tool

Group Participation Evaluation Forms I and II

One of the challenges to assessing student who work in collaborative teams is assessing group participations. Students need to know that each group member must pull his or her weight. As a component of a complete assessment system, especially in a collaborative learning environment, it is often helpful to engage students in a self-assessment of their participation in the group. Knowing that their contributions to the group will be evaluated provides an additional motivational tool to keep students constructively engaged.

Group Participation Evaluation Forms I and II provides students with an opportunity to assess group participation. In no case should the results of this evaluation be used as the sole source of assessment data. Rather, it is better to assign a weight to the results of this evaluation and factor it in with other sources of assessment data. If you have not done this before, you may be surprised to find how honestly students will critique their own work, often more intensely than you might do.

Investigation 3: The Relationship between the Earth and Its Moon

5. Take it one step further and think about this. The planet Jupiter has a gravitational pull that is about $2\frac{1}{2}$ (2.5) times greater than that of the Earth. How far could each of you jump on Jupiter?

   a) Do the calculations and place this information in the data table.

6. Measure these distances on the floor and make tape markers for this as well. Label each JJ (Jupiter Jump).

Mini-Investigation C: Our Moon and the Earth's Tides

In this last part of the GravLab, you will pull together what you know about the Moon and gravity to explore a major effect that the Moon has on the Earth.

1. Look at the tide table on the following pages. This table shows the level of the ocean tides during August 2004. It also shows the phases of the Moon for that same month.

2. Review what you learned in the previous investigation about the phases of the Moon and what you now know about gravity. Now, look carefully at the data in the table. Talk about the following questions. First talk with your group and then with the rest of the class. Then answer the questions in your journal.

   a) What do you think is the relationship (if any) between the tides and the phases of the Moon?

   b) How might you explain this relationship knowing what you know about the Earth, the Moon, and the pull of gravity?

Sharing and Discussing Your Findings

1. When you finish all of the GravLab, summarize what your group has learned about gravity in several clear and complete sentences.

   a) Record these sentences in your journal.

2. When you have written them in your journal, compare them to the whole-class definition of gravity.

   a) Do you need to change anything about the class definition, or add to it? Work as a class, with the help of your teacher, to come up with as complete and accurate a definition as possible. Record this in your jounal.

Inquiry

Using Data Tables as Scientific Tools

Scientists collect and review data using tools. You may think of tools as only physical objects like telescopes and measuring tapes. However, forms in which information is gathered, stored, and presented are also tools for scientists. In this investigation you are using a tide table as a scientific tool.

Teacher Commentary

5. Once again, students will need to draw upon their earlier understanding of the relative force of gravity on the Moon and on Earth to be able to calculate the different jumps.

Mini-Investigation C: Our Moon and the Earth's Tides
2. a) Students should see that the tides are greatest during the full or new Moon phases.
   b) Answers will vary but students should understand that when the Moon, Sun, and the Earth are lined up, as during a full or new Moon, then the pull of gravity is going to be the greatest, creating the biggest tides. During a quarter Moon, the Moon and the Sun are pulling at opposite directions on the Earth, which will make the tides lower.

INVESTIGATING EARTH IN SPACE: ASTRONOMY

Ocean City (fishing pier), Maryland 38.3267° N, 75.0833° W

August 2004

Day	High	Low	High	Low	High	Moon	Sunrise	Sunset
Sun 01		02:41 / -0.25 ft	08:29 / 3.50 ft	14:36 / -0.58 ft	21:03 / 4.75 ft		06:02	20:09
Mon 02		03:29 / -0.29 ft	09:22 / 3.61 ft	15:29 / -0.50 ft	21:52 / 4.53 ft		06:03	20:08
Tue 03		04:15 / -0.24 ft	10:13 / 3.68 ft	16:21 / -0.31 ft	22:39 / 4.24 ft		06:04	20:07
Wed 04		05:01 / -0.13 ft	11:02 / 3.69 ft	17:15 / -0.06 ft	23:25 / 3.87 ft		06:05	20:06
Thu 05		05:46 / 0.03 ft	11:51 / 3.65 ft	18:10 / 0.23 ft			06:06	20:05
Fri 06	00:10 / 3.48 ft	06:32 / 0.21 ft	12:42 / 3.58 ft	19:07 / 0.50 ft			06:07	20:04
Sat 07	00:58 / 3.12 ft	07:18 / 0.40 ft	13:35 / 3.50 ft	20:06 / 0.72 ft		Last Quarter	06:08	20:03
Sun 08	01:49 / 2.81 ft	08:05 / 0.57 ft	14:31 / 3.45 ft	21:06 / 0.88 ft			06:09	20:02
Mon 09	02:46 / 2.61 ft	08:54 / 0.68 ft	15:31 / 3.47 ft	22:07 / 0.96 ft			06:09	20:00
Tue 10	03:45 / 2.54 ft	09:46 / 0.73 ft	16:28 / 3.54 ft	23:08 / 0.96 ft			06:10	19:59
Wed 11	04:41 / 2.58 ft	10:39 / 0.70 ft	17:20 / 3.67 ft				06:11	19:58
Thu 12		00:00 / 0.90 ft	05:32 / 2.69 ft	11:31 / 0.61 ft	18:07 / 3.83 ft		06:12	19:57
Fri 13		00:43 / 0.79 ft	06:18 / 2.85 ft	12:20 / 0.48 ft	18:50 / 3.98 ft		06:13	19:56
Sat 14		01:21 / 0.67 ft	07:01 / 3.03 ft	13:04 / 0.34 ft	19:32 / 4.12 ft		06:14	19:54

Investigation 3: The Relationship between the Earth and Its Moon

Ocean City (fishing pier), Maryland 38.3267° N, 75.0833° W							
August 2004							
Sun 15		01:58 / 0.45 ft	07:44 / 3.21 ft	13:47 / 0.24 ft	29:12 / 4.21 ft	New Moon	06:15	19:53
Mon 16		02:33 / 0.42 ft	08:25 / 3.38 ft	14:28 / 0.17 ft	20:51 / 4.25 ft		06:16	19:52
Tue 17		03:08 / 0.33 ft	09:07 / 3.54 ft	15:09 / 0.16 ft	21:29 / 4.21 ft		06:16	19:50
Wed 18		03:45 / 0.28 ft	09:47 / 3.69 ft	15:52 / 0.20 ft	22:08 / 4.11 ft		06:17	19:49
Thu 19		04:22 / 0.26 ft	10:29 / 3.83 ft	16:38 / 0.29 ft	22:47 / 3.93 ft		06:18	19:48
Fri 20		05:02 / 0.28 ft	11:12 / 3.94 ft	17:28 / 0.40 ft	23:30 / 3.69 ft		06:19	19:46
Sat 21		05:45 / 0.31 ft	11:59 / 4.01 ft	18:35 / 0.52 ft			06:20	19:45
Sun 22	00:16 / 3.43 ft	06:32 / 0.36 ft	12:52 / 4.05 ft	19:23 / 0.63 ft			06:21	19:44
Mon 23	01:09 / 3.17 ft	07:25 / 0.38 ft	13:52 / 4.08 ft	20:28 / 0.69 ft		First Quarter	06:22	19:42
Tue 24	02:11 / 2.98 ft	08:24 / 0.37 ft	14:59 / 4.15 ft	21:35 / 0.67 ft			06:23	19:41
Wed 25	03:19 / 2.90 ft	09:27 / 0.30 ft	16:07 / 4.27 ft	22:43 / 0.57 ft			06:24	19:39
Thu 26	04:27 / 2.97 ft	10:33 / 0.16 ft	17:11 / 4.42 ft	23:47 / 0.39 ft			06:24	19:38
Fri 27	05:30 / 3.157 ft	11:37 / -0.04 ft	18:10 / 4.56 ft				06:25	19:36
Sat 28		00:43 / 0.18 ft	06:27 / 3.38 ft	12:37 / -0.23 ft	19:04 / 4.63 ft		06:26	19:35
Sun 29		01:33 / -0.00 ft	07:21 / 3.61 ft	13:32 / -0.37 ft	19:54 / 4.62 ft	Full Moon	06:27	19:33
Mon 30		02:18 / -0.13 ft	08:12 / 3.82 ft	14:23 / -0.41 ft	20:41 / 4.50 ft		06:28	19:32
Tue 31		03:01 / -0.17 ft	09:00 / 3.97 ft	15:12 / -0.33 ft	21:26 / 4.29 ft		06:29	19:30

INVESTIGATING EARTH IN SPACE: ASTRONOMY

Digging Deeper

As You Read...
Think about:
1. What does gravitational attraction depend on?
2. Why doesn't the Sun's gravity pull all the objects in our Solar System into the Sun?
3. How does the Moon affect the Earth's tides?

GRAVITY HOLDS THE SOLAR SYSTEM TOGETHER

It is impossible to talk about the relationships among the Earth, Moon, and Sun without first learning about gravity. Gravity is the attraction of all objects to all other objects. Gravitational attraction depends on the mass (amount of matter) of the objects and the distance between them. The greater the mass of the objects, the greater is the force. The greater the distance between the objects, the smaller is the force. You can't see gravity. As a force, it is invisible.

The Sun is very far away from the Earth (approximately 150 million kilometers), but it has tremendous mass. Because of its incredible mass, the Sun has the strongest gravity in the Solar System. As the planets orbit the Sun, gravity pulls them towards the center of the Sun. Although this pull exists, the planets do not move towards the Sun's center. This is because of their forward motion. The forward movement of each planet is balanced against the Sun's gravitational pull. As a result, the planets move constantly through their orbit just fast enough to stop them from being pulled toward the Sun.

The Sun's mass is so great that its gravity keeps the Earth (and all the other objects in the Solar System) orbiting around it. The Moon is much smaller in mass than the Earth.

Like the planets around the Sun, the Moon's orbit around the Earth is controlled by its forward movement and the gravitational pull of the Earth.

Teacher Commentary

Digging Deeper

The Lunar Cycle

This section provides text, illustrations, and diagrams that give students greater insight into the relationship between the Earth and the Moon. You may wish to assign the **As You Read** questions as homework to help students focus on the major ideas in the text.

As You Read…

Think about:

1. Gravitational attraction between two or more objects depends on the mass of the objects and the distance between them.

2. The objects in the Solar System are not pulled into the Sun because although the Sun does pull them, they also have a forward motion. This forward motion is balanced against the Sun's gravitational pull.

3. The Moon causes the Earth's water to move toward it slightly, creating the tides.

This area is a complex set of scientific concepts. Initially, it may be helpful to divide this **Digging Deeper** into sections and have different groups of students specialize in them. However, it will be important that everyone has an opportunity to understand the full picture.

Be sure to allow time for a full discussion after students study the **Digging Deeper** section. Be alert for any student misunderstandings and find ways to clarify the concepts involved.

Assessment Opportunity

You may wish to rephrase selected questions from the **As You Read** section into multiple choice or true/false format to use as a quiz. Use this quiz to assess student understanding and as a motivational tool to ensure that students complete the reading assignment and comprehend the main ideas.

Investigation 3: The Relationship between the Earth and Its Moon

However, it is relatively close to Earth and revolves around Earth. Just as the Earth's gravity keeps the Moon traveling around it, the Moon's gravity causes certain events on the Earth.

THE RISE AND FALL OF THE TIDES

The gravitational pull of the Moon is strong enough to cause the Earth's oceans to move slightly towards it. The ocean's rise in height forms a *tidal bulge* as water moves towards the Moon.

If part of the Earth's surface within the bulge has a coastline, then it will experience a *high tide*. A high tide also occurs on the opposite side of the Earth at the same time. During the Earth's 24-hour rotation, part of a coast can move into a tidal bulge. It will have a high tide inside the bulge. Then a *low tide* occurs after a coast has moved away from the bulge. Each coastal location has two high tides and two low tides daily. Together these make one *tidal cycle*.

The Sun's gravitational pull also affects the Earth's oceans. The Sun pulls the Earth's oceans in the same way as the Moon. When the Sun, the Moon, and the Earth are in alignment (during a new Moon or a full Moon), the highest high tides and the lowest low tides occur. During the half-Moon phases, the Sun and Moon pull the oceans in different directions. At this time, high tides aren't as high and low tides aren't as low.

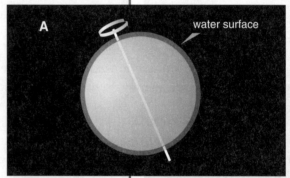

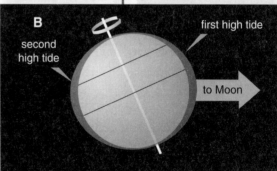

The gravitational pull of the Moon causes the Earth's oceans to form a tidal bulge.

Teacher Commentary

NOTES

INVESTIGATING EARTH IN SPACE: ASTRONOMY

The left image shows high tide and the right image shows low tide.

A SYSTEM IN MOTION

Despite the gigantic size of the Solar System, some of the most important processes that affect the Earth can be predicted. This is because different parts of the Solar System, such as moons or planets, move with regular cycles. Just as the Earth is a system with many parts, it is also part of a larger system involving the Sun and the Moon. Each planet and moon is also part of a larger system, called the Solar System. The key to the Solar System is the gravitational force that ties it all together.

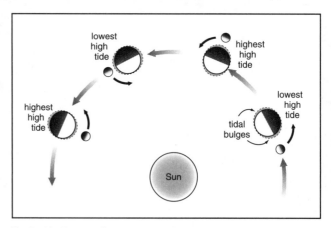

The Earth's tides are affected by the gravitational pull of both the Moon and the Sun.

Teacher Commentary

> **About the Photo**
> The photos on page A24 of the student text shows high and low tide in the same area. These photos were probably taken approximately 6 hours apart.

Investigation 3: The Relationship between the Earth and Its Moon

Review and Reflect

Review

1. How does gravity affect our Solar System?
2. What factors affect gravity?
3. What effect does the Moon have on the Earth?
4. What effect does the Earth have on the Moon?

Reflect

5. How might motion on the Earth be different if the Earth was half its mass? Why do you think that?
6. How might motion on the Earth be different if the Earth was twice its mass? What reason do you have for that?

Thinking about the Earth System

7. How does the Moon's and Earth's gravity affect the hydrosphere?
8. How does this effect on the hydrosphere affect the other major systems. Remember to write connections, as you find them, on the *Earth System Connection* sheet.

Thinking about Scientific Inquiry

9. In which parts of the investigation did you:
   a) Make a prediction?
   b) Use tools to measure?
   c) Record your own ideas?
   d) Revise your ideas?
   e) Compare data to look for patterns and relationships?
   f) Share ideas with others?
   g) Find information from different sources?
   h) Pull your information together to make a presentation?

Teacher Commentary

Review and Reflect

Review

Give your students ample time to review what they have learned in **Investigation 3**. In student answers, look for evidence of an understanding of the processes involved, as well as for any misconceptions that still remain. Encourage students to express their ideas clearly and to use correct science terminology where appropriate.

1. Gravity is what keeps the Solar System together and keeps the objects in the Solar System in their current locations.

2. Gravity is affected by the size of the objects and the distances between them.

3. The Moon has a gravitational pull that pulls the waters of the Earth. This creates the tides.

4. The Earth has a gravitational pull on the Moon which keeps it in orbit around our planet.

Reflect

Give students time to reflect on the nature of the evidence they generated in their investigations. Again, help them to see that evidence is crucial in scientific inquiry.

5. Answers will vary.

6. Answers will vary.

Thinking about the Earth System

7. • The Moon's gravitational pull is strong enough to cause Earth's oceans (hydrosphere) to move slightly towards it.
   • The oceans' rise in height forms a tidal bulge as water is attracted by the Moon's gravity.

8. • The rising and falling of ocean tides affects shorelines in the geosphere causing erosion and also the deposition of sediments.
   • Tidal flow has an impact on aquatic life forms and their environments.
   • Humans have to adapt their shipping and recreational activities to the rise and fall of tides.
   • Tidal energy can be harnessed to generate electrical energy.

Thinking about Scientific Inquiry

In **Investigation 3,** your students have been using data tables and modeling experiments to clarify the relationship between the Earth and the Moon. Have your students carefully consider the list of **Inquiry Processes** shown on page x of their edition and record which of them they have used in this investigation. (A **Blackline Master Astronomy 1.2,** is included in the Appendix.)

Teacher Review

Use this section to reflect on and review the investigation. Keep in mind that your notes here are likely to be especially helpful when you teach this investigation again. Questions listed here are examples only.

Student Achievement

What evidence do you have that all students have met the science content objectives?
_____

Are there any students who need more help in reaching these objectives? If so, how can you provide this? _____
_____

What evidence do you have that all students have demonstrated their understanding of the inquiry processes? _____
_____

Which of these inquiry objectives do your students need to improve upon in future investigations? _____
_____

What evidence do the journal entries contain about what your students learned from this investigation? _____
_____

Planning

How well did this investigation fit into your class time? _____
_____

What changes can you make to improve your planning next time? _____
_____

Guiding and Facilitating Learning

How well did you focus and support inquiry while interacting with students?
_____

What changes can you make to improve classroom management for the next investigation or the next time you teach this investigation? _____
_____

Teacher Review

How successful were you in encouraging all students to participate fully in science learning? _____

_____

How did you encourage and model the skills values, and attitudes of scientific inquiry? _____

_____

How did you nurture collaboration among students? _____

_____

Materials and Resources

What challenges did you encounter obtaining or using materials and/or resources needed for the activity? _____

_____

What changes can you make to better obtain and better manage materials and resources next time? _____

_____

Student Evaluation

Describe how you evaluated student progress. What worked well? What needs to be improved? _____

_____

How will you adapt your evaluation methods for next time? _____

_____

Describe how you guided students in self-assessment. _____

_____

Self Evaluation

How would you rate your teaching of this investigation? _____

_____

What advice would you give to a colleague who is planning to teach this investigation? _____

_____

Investigating Earth in Space: Astronomy – Investigation 3

NOTES

INVESTIGATION 4: FINDING OUR PLACE IN SPACE

Background Information

The closest star to the Earth is the Sun. The average distance between the Earth and the Sun is an average 150 million kilometers. This distance is defined as an astronomical unit or one AU.

The next closest star to the Earth is Alpha Centauri. Alpha Centauri is about 40 trillion kilometers away! This is nearly 300,000 times as far away as the Sun is from the Earth. Kilometers are not an efficient way to express distances of that magnitude. And AU is not a reasonable way to express it either. Instead, scientists describe distances of this magnitude using light-years. A light-year is a measurement of distance not time. A light-year is the distance a light ray can travel in one year. This speed is about 300,000 kilometers per second. At this rate, light can travel about 9.5 trillion kilometers in 1 year. Alpha Centauri is about 4.3 light-years away from Earth. Betelgeuse, a supergiant in the constellation Orion, is almost 490 light-years away.

To the unaided eye, many of the planets in the Solar System look like stars. However, this is one difference that the earliest of sky observers noticed. Over time, the position of the stars does not change nautically with respect to each other. This is why the constellations are such recognizable patterns. The position of the planets, however, changes constantly. The reason for this is because the distances between Earth and the two objects. The Earth is much closer to the other planets so they appear to move while the stars are much farther away.

Early scientists described the Solar System as being heliocentric, that is, with the Earth in the center. Ptolemy, a Greek astronomer, developed an Earth-centered model of the Solar System in which he predicted the locations of the planets. This model was accepted until the 1600s.

Copernicus is credited with developing the heliocentric model of the Solar System. He suggested that the Earth and the rest of the planets revolved around the Sun. Retrograde motion would occur whenever the Earth passed another planet. This proved to be a much simpler explanation of the observed motions in the sky.

More Information…on the Web

Visit the *Investigating Earth Systems* web site www.agiweb.org/ies/ for links to a variety of web sites that will help you deepen your understanding of content and prepare you to teach this investigation.

Investigation Overview

In **Investigation 4**, students become familiar with the planets within our Solor System. They become experts on one of the nine planets by creating a scale model of the Solar System. The **Digging Deeper** section reviews the objects visible from the Earth at night as well as the way in which scientists measure distances in space.

Students use a variety of resources on the Solar System, including: the **Digging Deeper** section, reference books, their own drawings, CD-ROMs, and web sites, if available, to learn about where the Earth is in space. They then make a model of the Solar System that they can display in the school's hallways.

Goals and Objectives

As a result of this investigation, students will locate the Earth in the Solar System and understand the relative distances between the Earth, the Sun, and the other planets in the Solar System.

Science Content Objectives

Students will collect evidence to verify that:

1. The Earth is the third planet from the Sun.
2. Some planets are larger than the Earth and others are smaller.
3. The planets all orbit the Sun and are different distances from the Sun and each other.

Inquiry Process Skills

Students will:

1. Use a variety of investigative resources.
2. Make a model of what they cannot see.
3. Revise models on the basis of additional observations.
4. Communicate observations and findings to others.

Teacher Commentary

Connections to Standards and Benchmarks

In **Investigation 4,** students take their initial ideas about the Solar System, compare it to reliable information and data sources, and then construct a scale model representation which is displayed for others to see and understand. The evidence they review and the model they construct will set the stage for understanding the wider Solar System. These observations and investigations contribute to developing the National Science Education Standards and AAAS Benchmarks below:

NSES Links

- Evidence consists of observations and data on which to base scientific explanations.
- The Earth is the third planet from the Sun in a system that includes the Moon, the Sun, eight other planets and their moons, and smaller objects, such as asteroids and comets. The Sun, an average star, is the central and largest body in the Solar System.
- Most objects in the Solar System are in regular and predictable motion. Those motions explain such phenomena as the day, the year, phases of the Moon, and eclipses.
- Gravity is the force that keeps planets in orbit around the Sun and governs the rest of the motion in the Solar System. Gravity alone holds us to the Earth's surface and explains the phenomena of the tides.

AAAS Links

- We live on a relatively small planet, the third from the Sun in the only system of planets definitely known to exist (although other, similar systems may be discovered in the universe).
- Everything on or anywhere near the Earth is pulled toward the Earth's center by gravitational force.
- Nine planets of very different size, composition, and surface features move around the Sun in nearly circular orbits. Some planets have a great variety of moons and even flat rings of rock and ice particles orbiting around them. Some of these planets and moons show evidence of geologic activity. The Earth is orbited by one Moon, many artificial satellites, and debris.
- Large numbers of chunks of rock orbit the Sun. Some of those that the Earth meets in its yearly orbit around the Sun glow and disintegrate from friction as they plunge through the atmosphere—and sometimes impact the ground. Other chunks of rocks mixed with ice have long, off-center orbits that carry them close to the Sun, where the Sun's radiation (of light and particles) boils off frozen material from their surfaces and pushes it into a long, illuminated tail.

Preparation and Materials Needed

Preparation

In **Investigation 4**, your students will be drawing a scale model of the Solar System using information they already knew about the Solar System and information and data found in the **Digging Deeper** section of the text.

This investigation requires five or six 40-minute class periods to complete, depending upon how you structure it. **Day One:** Have the students address the **Key Question** and review the kinds of ideas they have and record them in their Journals.

A Solar System diagram is included in the Student Edition. You might find it useful to use a computer projector to show the CD-ROM on the Solar System to all your students at once. You will also need to arrange for students to post up their Solar System models in the school hallways, or other similar space.

This investigation should fit into three 40-minure class periods. **Day One:** Have students think about and discuss the **Key Question**, review the ideas they have and record them in their Journals. Students should then be able to complete items 1 – 3. **Day Two:** Have students continue through items 4 – 6. **Day Three:** Have students share their findings and **Review and Reflect** on the whole investigation.

Suggested Materials

For this investigation, your group will need:
- sketch of Solar System from the Pre-Assessment
- drawing materials (colored pencils or markers and paper)
- diagram of Solar System
- CD-ROM or video on the Solar System (if possible)
- computer to view CD-ROM or video player and monitor
- long narrow roll of paper (such as adding machine paper roll)
- masking tape
- metric measuring tape
- access to a school hallway
- calculators

Teacher Commentary

NOTES

INVESTIGATING EARTH IN SPACE: ASTRONOMY

Investigation 4:
Finding Our Place in Space

Key Question
Before you begin, first think about this key question.

Where is the Earth in space and how do scientists know?

Materials Needed

For this investigation your group will need:

- drawing materials (colored pencils or markers and paper)
- diagram of Solar System
- long narrow roll of paper
- masking tape
- metric measuring tape
- access to a school hallway
- calculator

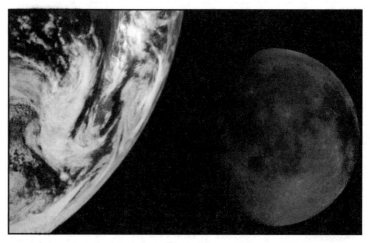

Think about what you know about the Earth in space. Share your thinking with others in your group and in your class. Keep a record of the discussion in your journal.

Share your group's ideas with the rest of the class.

Investigate
In the first **Investigation**, you learned some of the characteristics of Earth. In this **Investigation**, you will learn about where the Earth is in space and how people came to discover this over time. You might be surprised to see how ideas about the Earth's position in our Solar System have changed.

Teacher Commentary

Key Question

Begin by asking students to respond to the **Key Question**, "Where is the Earth in space and how do scientists know?" Tell students to write down their ideas in their journals. After a few minutes, discuss students' ideas in a brief conversation. Emphasize thinking and sharing of ideas. Avoid seeking closure (i.e., the "right answer"). Record all of the ideas that students share on an overhead transparency or on the board. Have students record this information in their journals.

Student Conceptions about the Earth in Space

The most common answer to the **Key Question** will probably be that Earth floats in space with the Moon close to the Sun. Some may mention other planets in our Solar System such as Mars, Venus, and Jupiter. Much will depend upon experiences students have had either with elementary school science or informal education. Do not be surprised if you get a wide range of responses with variable accuracy. Some students may know something about the history of astronomy, and many may think that our knowledge of Earth and Space has been recently acquired through NASA and modern space exploration. They are unlikely to have heard of Galileo or any of the other historical astronomers.

Assessment Tool

Key Question-Evaluation Sheet

Use the evaluation sheet to help students understand and internalize basic expectations for the starting activity. The **Key Question-Evaluation Sheet** emphasizes that you want to see evidence of prior knowledge and that students should communicate their thinking clearly. You will not be likely to have time to apply this assessment every time students complete a starting activity; yet, in order to ensure that students value committing their initial conceptions to paper and taking the activity seriously, you should always remind them of the criteria. When time permits, use this evaluation sheet as a check on the quality of their work. As with any assessment tool used in *IES*, the instrument should be provided to students and discussed *before* they complete a task. This ensures that they have a clear understanding of your expectations for their work.

Assessment Tool

Journal Entry-Evaluation Sheet

A photocopy master of this tool is included in the Appendix. Remind students that his sheet provides general guidelines for assessing student journals. Adapt this sheet so that it is appropriate for your classroom. Remind students what they are expected to include in their journals and how their entries will be assessed. Use this tool to: (a) provide a record of student progress for your own use, and (b) give feedback to students, by giving them a copy of your assessment.

> **Assessment Tool**
> Journal Entry-Evaluation Sheet
> Use this checklist as a guide for quickly checking the quality and completeness of journal entries.

Answer for the Teacher Only

Earth is within a galaxy known as the Milky Way, which is a large spiral galaxy. Earth is located in one of the spiral arms of the Milky Way (called the Orion Arm), which lies about two-thirds of the way out from the center of the galaxy. Planet Earth is part of the Solar System. It has nine planets and numerous asteroids, all of which orbit the Sun. Earth is the third planet from the Sun in the Solar System. Humans have determined this by an evolving process of observation, hypothesis and theory, starting with a Flat-Earth Geocentric Universe, to a Heliocentric Universe, to a Milky Way Centered Universe, to a Universe with no center.

Teacher Commentary

NOTES

Investigating Earth in Space: Astronomy

Investigation 4: Finding Our Place in Space

1. Without looking ahead, draw the Sun and the planets and the distances from the Sun to the planets as nearly to scale as you can. Compare your drawing to the diagram of the Solar System in this book.

   a) Make a list of all the differences you observe between your drawing and the diagram.

2. In this **Investigation** you will be making a scale model of the Solar System. To do this you need to be able to answer the following questions. You may use the information in the **Digging Deeper** section and the table on the next page to help you answer these questions.

   a) What is at the center of our Solar System?

   b) What are the names of all the planets in our Solar System?

   c) Where is the Earth in the Solar System?

   d) How far is the Earth's Moon from the Earth?

   e) How far is the Earth from the Sun and the other planets in the Solar System?

3. The objects in the Solar System are great distances away from one another. How can they be a "system"? A system is made up of parts that depend upon each other. Think of the cars, roads and bridges of a transportation system or the wires, signals, and receivers of a communication system.

   a) Can objects so far away from each other in space really be dependent on one another? In what ways?

4. You will now design and construct a model of the Solar System. The table shown on the next page will help you with the relative sizes of the Sun and planets and their average distances from the Sun. Your goal is to be able to make a model of the Solar System that you can place in the hallways in your school. This will give you enough space to be able to model the relative distances between the planets. You still may have some trouble modeling the different sizes of the planets and the Sun. Do the best you can to get the distances as much to scale as possible, even if you can't do the sizes of the planets.

5. Use this scale to make your model:
   1 cm (centimeter) = 1,000,000 km (kilometers).

Teacher Commentary

Investigate

Teaching Suggestions and Sample Answers

2. a) The Sun is at the center of our Solar System.
   b) The planets are: Mercury, Venus, Earth, Mars, Jupiter, Saturn, Uranus, Neptune, and Pluto.
   c) The Earth is in an orbit around the Sun and is the third planet from the Sun (between Venus and Mars).
   d) The distance between the Moon and the Earth varies between 363,300 km and 405,500 km. The mean distance is 384,400 km.
   e) The average distance between the Earth and the Sun is 149,597,890 km.

3. a) Yes, objects far from one another in space can be dependent upon each other. For example, these objects will often experience the pull of gravity from a distant object.

Managing Scale

Scale can be a difficult concept for students, especially when the distances are so large. Modeling the Solar System to scale is quite a formidable task in terms of the mathematics involved. You will need to check on your students' abilities to understand and use scale. This might be a good opportunity to collaborate with your students' mathematics teacher(s). He or she might be able to devote some class time to scale, or help you introduce your students to it in science.

Because of the scale, you will need a large area to display the model. The hallways around the school provide one solution. Another might be to use the walls of the cafeteria, gymnasium, or auditorium.

Drawing the Planets

It would be good to draw these planets to scale where possible. One easy way could be to take an existing to-scale picture of diagram of the planets and cut them out for mounting, or have students trace from them. Keep in mind that the Sun is especially large, and Pluto is very small in relation to the other planets.

INVESTIGATING EARTH IN SPACE: ASTRONOMY

Inquiry
Using Mathematics

Scientists often use mathematics in their investigations. In this investigation you used a scale to construct your model of the Solar System.

Table 1: Diameters of the Sun and Planets and Distances from the Sun							
Planet	Diameter (km)	Kilometers from Sun (multiplied by 1,000,000)	Astronomical Unit	Average Temperature (Degrees Celsius)	Number of Moons	Orbital Period (Earth Days)
Mercury	4879	57.9	0.39	167	0	88.0
Venus	12,104	108.2	0.72	464	0	224.7
Earth	12,756	149.6	1.0	15	1	365.2
Mars	6794	227.9	1.52	-65	2	687.0
Jupiter	142,984	778.6	5.20	-110	61*	4331
Saturn	120,536	1433.5	9.54	-140	31*	10,747
Uranus	51,118	2872.5	19.19	-195	26*	30,589
Neptune	49,528	4495.1	30.07	-200	13*	59,800
Pluto	2390	5870.0	39.48	-225	1	90,588

Work in specialist groups in the class to draw the Sun and the planets. For example, one group might draw the Earth, another group Jupiter, etc.

6. When all the planets and Sun are complete, tape them in the hallway, adding labels with the planets' names. If you have a long roll of paper (such as adding-machine paper), you could put this on the wall from planet to planet, writing the distances between the planets on it with markers.

Inquiry
Modeling

To investigate the great distances between objects in the Solar System, you will be making a model. Models are very useful scientific tools. Scientists use models to simulate real-world events. Since you cannot travel from one planet to the next to discover how far apart they are, you will make a model that will let you make the journey in the hallway of your school.

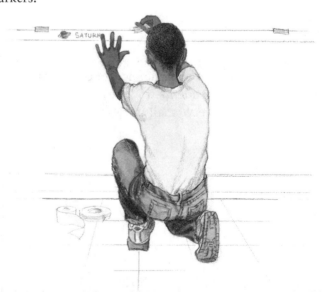

Teacher Commentary

NOTES

Investigation 4: Finding Our Place in Space

Digging Deeper

THE SKY AT NIGHT

When you look up into the sky at night, what do you see? On a clear night, you might see some of the billions of stars that make up the *Milky Way Galaxy*. Light from these stars travels many hundreds of trillions of kilometers before it reaches your eyes.

The Moon is the closest object that you might see in the night sky. The Moon is approximately 384,403 km from Earth. Compared to other objects in the Solar System, this is not very far away. At certain times, the Moon appears quite large and bright. If you look carefully, you may notice some of the Moon's surface features, such as craters and maria. Maria are the dark flat areas covered by the black rock *basalt*.

You might also see some planetary neighbors. The nearest planet to the Earth that you might see is Venus. This planet can be as close as 38 million kilometers away when it passes Earth in its orbit, or as far away as 261 million kilometers! Mars is the next closest planet to the Earth with a distance that varies from 56 million kilometers to about 401 million kilometers.

In between the stars, the Moon, and the planets you would see blackness. This is space that fills the

When you look up into the sky at night, you can see planets and some of the billions of stars that make up the Milky Way Galaxy.

As You Read...
Think about:
1. What are some of the objects you can expect to see in the sky on a clear night?
2. How do astronomers measure distance between objects in the Solar System?
3. How do astronomers measure distance between objects outside the Solar System?
4. What are the names of the planets in the Solar System and where are they in relation to the Sun?

Teacher Commentary

Digging Deeper

This **Digging Deeper** shows students where Earth is located within the Solar System, Milky Way, galaxy and universe. It explains how gravity and the forward motion of orbiting objects in space hold systems together.

Assign the **Digging Deeper** reading section and **As You Read** questions to students. Later you can discuss their answers as a class.

As You Read...

Think about:
1. On a clear night, you might expect to see planets, the Moon, stars, constellations, satellites orbiting the Earth, and part of the Milky Way.
2. Distances within the Solar System are measured in astronomical units, or AU. One AU is the average distance between the Earth and the Sun.
3. Distances outside our Solar System are measured in light-years. A light-year is the distance light travels in one year. This is equal to 9.5×10^{12} km.
4. The Solar System is composed of the Sun, the four inner Solar System planets: Mercury, Venus, Earth, Mars; the Main Asteroid Belt; the outer Solar System planets Jupiter, Saturn, Uranus, Neptune, and Pluto; more than 147 satellites of the planets, including moons; and smaller bodies such as comets, meteoroids and asteroids.

INVESTIGATING EARTH IN SPACE: ASTRONOMY

The Moon is a familiar object in the night sky. The dark areas are maria.

huge gaps between objects. It contains very little matter, except for a small amount of dust that is usually too small and far away to be seen.

EARTH'S NEAREST STELLAR NEIGHBORS

How far does light from the stars in the sky travel to reach the Earth? The Sun is the closest star to the Earth and is about 150 million kilometers away. The second nearest stellar neighbor is Proxima Centauri. Its light travels 40,000,000,000,000 km to reach the Earth!

The distances between objects in space are so huge that Earth-based units are not very useful measurements. Instead, scientists use distance units of measure that are more appropriate. For distances within the Solar System, astronomers use the astronomical unit (AU). One AU is the average distance between the Earth and the Sun or about 150 million kilometers. For distances between stars, astronomers use light-years. Light travels about 300,000 km per second. In one year it can travel 9,460,000,000,000 km. Light traveling from Proxima Centauri to the Earth takes 4.22 years. Using this measurement, you can say that Proxima Centauri is 4.22 light-years (LY) from Earth.

MODELING THE SOLAR SYSTEM

Scientists often use models when they examine the moons, planets, stars and other objects in space. A model can be a small object that represents a larger object. Since the sizes of stars and planets are so great, scientists build small models of space systems.

A composite picture of the planets in the Solar System.

These models give scientists a clearer idea of what objects are in space and how those objects interact. To

Teacher Commentary

Space Measurements

The concept of time is one of the most difficult things to understand, especially when measured over vast periods, such as geologic time. When it comes to objects in the universe, the huge distances between them are almost impossible to comprehend and cannot be sensibly expressed using normal linear measurement units such as kilometers or miles. Instead "light-years" are used using the formula shown: 9,500,000,000,000 kilometers per year, which is the approximate distance light travels in one year. This is a complex idea for your students to grasp. Help them to understand some of the mathematics involved.

Planet Data Table

Encourage students to study this data table carefully. Have them look for the largest and smallest planets first and compare their diameters. They can then compare these dimensions with that of Earth and the other planets. This is also a good opportunity for students to learn more about interpreting data tables. Ask students to look for any patterns and relationships they can see in these data and write them down. Also ask them to list any questions that may arise. For example, some students may be surprised to learn that Jupiter has so many moons and want to find an explanation for this.

Investigation 4: Finding Our Place in Space

build a model, scientists must create objects and distances that are proportional to what is actually found in space. Because of the huge distances in space and the relatively small diameters of planets, making a scale model of the Solar System can take up a lot of space!

THE EARTH'S POSITION IN SPACE

The position of the Earth in space depends on its location in its orbit around the Sun. An orbit is the path and motion of one object around another object. If you look down on the Solar System from the North, with respect to Earth, each planet in the Solar System revolves around the Sun in a counter-clockwise direction. This is due to the gravitational force between the Sun and the planets. The relative position of the planets to one another depends on their positions in their orbits.

Another way to think of orbits is as a large racetrack with parallel lanes. The Sun lies near the center of the flat surface of the track. Planets race along in their lanes at different speeds and over different distances. The planets closer to the Sun on the inside lanes travel faster and over a shorter distance than the planets in lanes farther away from the Sun.

Can you imagine living on a planet that takes 248 years to orbit the Sun? This would be the case if you lived on Pluto. If you lived on Mercury, however, the closest planet to the Sun, your orbit would only take 88 days! On Earth, it takes us an average of 365.2 days to orbit the Sun.

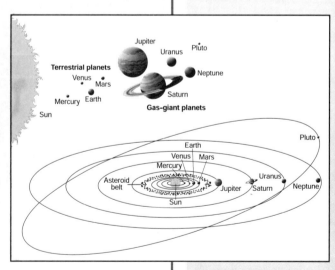

The orbits of the planets around the Sun.

Teacher Commentary

Modeling the Solar System

One way to help visualize the relative sizes within the Solar System is to think of a model in which everything is reduced in size by a factor of a billion. This way the model of Earth would be about 1.3 cm in diameter (or the size of a grape). Earth's Moon would be about 37 cm (about 1.5 feet) from the Earth. The Sun would be 1.5 meters in diameter (about the height of a man) and 150 meters (that's about a city block) from the Earth. Jupiter would be 15 cm in diameter (the size of a large grapefruit) and 5 blocks away from the Sun. Saturn (the size of a big orange) would be 10 blocks away; Uranus and Neptune (tennis balls) 20 and 30 blocks away. A human on this scale would be the size of an atom. However, the nearest star would be over 40,000 km away. (Adapted from: http://www.nineplanets.org/)

The Earth's Position in Space

Understanding that the planets in our Solar System orbit the Sun is the key concept for students to grasp here. With this, they can then see more easily how different planets orbit the Sun in different, yet similar ways, and at different distances and orbital speeds. They need to appreciate that Earth's distance from the Sun gives it special properties that other planets in the Solar System do not have, especially its range of temperatures, which contribute to the planet's capacity to sustain life.

INVESTIGATING EARTH IN SPACE: ASTRONOMY

Review and Reflect

Review

1. What was the most difficult part in making the model of the Solar System ? Why was this so?
2. If you traveled at the rate of 300 km/h (kilometers per hour), how long would it take you to get to Mars? To Jupiter?

Reflect

3. In what ways is the Solar System a true "system"?
4. What relationship do all of the planets have to the Sun? To each other?

Thinking about the Earth System

5. How does understanding the Earth System help you understand the Solar System?

Thinking about Scientific Inquiry

6. In which parts of the investigation did you:
    a) Make a model?
    b) Use mathematics to solve a problem?
    c) Revise your ideas?
    d) Use your imagination?
    e) Share ideas with others?
    f) Find information from different sources?

Teacher Commentary

Review and Reflect

Help your students carefully consider what they have done and learned in this investigation. As you do so, look for evidence of their understanding. Also, be alert for misunderstandings and find ways to help students develop ideas that conform to scientifically accepted concepts and explanations. It is important that you allow a good amount of time for students to review and reflect. This part of each investigation should never become trivialized or routine. Careful thought and consideration here is essential for solid understanding on your students' part.

Review

1. The chosen scale will have been an important determinant of how the model was constructed. Since scale can be difficult to grasp, ensure that all students review this carefully.
   It is likely that the most difficult part of modeling the Solar System is the sheer size of the distances between the Sun and the planets. Another will have been the relative sizes of the bodies that make up the Solar System, especially the huge size of the Sun in relation to the smaller planets like Pluto. Getting them all "on the same page" is impossible; making a long wall display the only obvious option.

2. Traveling at the speed of light, or in light-years, is also a difficult concept to comprehend. Check that your students have developed at least a working understanding of this measurement. The mathematical processes and answers to this question will alert you to any problems. It would take 261,000 hours to get to Mars. It would take roughly 2,096,667 hours to get to Jupiter at that rate.

Reflect

3. This concept of "system" is something that you will need to revisit with your students many times during this module. The key is to have them focus upon the components that are interconnected, and consider how changes in one would affect the others. Note students' responses to this question and, if necessary, hold a class discussion about their answers and ideas. The students should leave with the understanding that it is the force of gravity, in particular the gravity created by the huge mass of the Sun, which supplies the interconnection that makes the Solar "System" a system.

4. The relationship here is that all Solar System planets orbit the Sun. This is because of their gravitational attraction in combination with their forward motion. Gravitational attraction also keeps the planets together within the system.

Thinking about the Earth System

5. Because Earth is one of nine (possibly ten) planets orbiting the Sun, understanding how Earth works as a system provides us with a tool for understanding how the Solar System works.

Thinking about Scientific Inquiry

6. Make sure students consider their use of **Inquiry Processes** carefully. They should check this off against the list shown on page x of the Student Edition. It is important that students are able to interpret their investigations in terms of these **Inquiry Processes**.

Assessment Opportunity
Review and Reflect Journal Entry-Evaluation Sheet
Depending upon whether you have students complete the work individually or within a group, use the **Review and Reflect** part of the investigation to assess individual or collective understandings about the concepts and inquiry processes explored. Whatever choice you make, this evaluation sheet provides you with a few general criteria for assessing content and thoroughness of student work. Adapt and modify the sheet to meet your needs. Consider involving students in selecting and modifying the assessment criteria.

Teacher Commentary

NOTES

Teacher Review

Use this section to reflect on and review the investigation. Keep in mind that your notes here are likely to be especially helpful when you teach this investigation again. Questions listed here are examples only.

Student Achievement
What evidence do you have that all students have met the science content objectives?
_____

Are there any students who need more help in reaching these objectives? If so, how can you provide this? _____
_____

What evidence do you have that all students have demonstrated their understanding of the inquiry processes? _____
_____

Which of these inquiry objectives do your students need to improve upon in future investigations? _____
_____

What evidence do the journal entries contain about what your students learned from this investigation? _____
_____

Planning
How well did this investigation fit into your class time? _____
_____

What changes can you make to improve your planning next time? _____
_____

Guiding and Facilitating Learning
How well did you focus and support inquiry while interacting with students?
_____

What changes can you make to improve classroom management for the next investigation or the next time you teach this investigation? _____
_____

Teacher Review

How successful were you in encouraging all students to participate fully in science learning? _____

_____

How did you encourage and model the skills values, and attitudes of scientific inquiry? _____

_____

How did you nurture collaboration among students? _____

_____

Materials and Resources

What challenges did you encounter obtaining or using materials and/or resources needed for the activity? _____

_____

What changes can you make to better obtain and better manage materials and resources next time? _____

_____

Student Evaluation

Describe how you evaluated student progress. What worked well? What needs to be improved? _____

_____

How will you adapt your evaluation methods for next time? _____

_____

Describe how you guided students in self-assessment. _____

_____

Self Evaluation

How would you rate your teaching of this investigation? _____

_____

What advice would you give to a colleague who is planning to teach this investigation? _____

_____

NOTES

Teacher Commentary

INVESTIGATION 5: THE SUN AND ITS CENTRAL ROLE IN OUR SOLAR SYSTEM

Background Information

The first day of summer in the Northern Hemisphere occurs on, or nearly on, June 21st of each year. This day has the longest period of daylight. Because the daily increase in the noon altitude of the Sun stops on this date, it is called the summer solstice. The Northern Hemisphere is at its maximum. The Sun is directly overhead at 23.5° north latitude. This is where the Tropic of Cancer is located. On this date, every spot on the Earth that is within 23.5° of the North Pole is having 24 hours of daylight. This is all the locations within the Arctic Circle. In the Southern Hemisphere, every point within the Antarctic Circle experiences 24 hours of darkness on June 21st.

After June 21st, the tilt of the Northern Hemisphere decreases as the Earth continues on its path around the Sun. As the tilt decreases, daylight hours in the Northern Hemisphere decrease while daylight hours in the Southern Hemisphere increase.

Winter begins in the Northern Hemisphere on or nearly on December 21st. This is the winter solstice, or the shortest day of the year. The Sun is directly over the Tropic of Capricorn which is located at 23.5° south latitude. There are two days each year when neither hemisphere is tilted toward the Sun. These days are between both solstices. On these dates, day and night are equal in length all over the Earth. These days are the spring equinox (March 21st) and the autumn equinox (September 23rd).

The Sun is the most prominent feature in our Solar System. It is the largest object and contains approximately 98% of the total Solar System mass. The Sun's outer visible layer is called the photosphere and has a temperature of 6000°C. This layer has a spotted appearance due to the turbulent eruptions of energy at the surface.

Solar energy is created deep within the core of the Sun. It is here that the temperature (15,000,000°C) and pressure (340 billion times Earth's air pressure at sea level) is so intense that nuclear reactions take place. This reaction causes four protons or hydrogen nuclei to fuse together to form one alpha particle or helium nucleus. The alpha particle is about 0.7 % less massive than the four protons. The difference in mass is expelled as energy and is carried to the surface of the Sun, through a process known as convection, where it is released as light and heat. Energy generated in the Sun's core takes a million years to reach its surface. Every second 700 million tons of hydrogen are converted into helium ashes. In the process 5 million tons of pure energy is released; therefore, as time goes on the Sun is becoming lighter.

The chromosphere is above the photosphere. Solar energy passes through this region on its way out from the center of the Sun. Solar flares are found in the chromosphere. Flares are bright filaments of hot gas emerging from sunspot regions. Sunspots are dark depressions on the photosphere with a typical temperature of 4000°C.

The corona is the outer part of the Sun's atmosphere. It is in this region that prominences appear. Prominences are immense clouds of glowing gas that erupt from the upper chromosphere. The outer region of the corona stretches far into space and consists of particles traveling slowly away from the Sun. The corona can only be seen during total solar eclipses.

The Sun appears to have been active for 4.6 billion years and has enough fuel to go on for another five billion years or so. So, the Sun is about halfway through its life. At the end of its life, the Sun will start to fuse helium into heavier elements and begin to swell up; ultimately growing so large that it will swallow the Earth. After a billion years as a red giant, it will suddenly collapse into a white dwarf — the final end product of a star like ours. It may take a trillion years to cool off completely.

More Information…on the Web
Visit the *Investigating Earth Systems* web site www.agiweb.org/ies for links to a variety of web sites that will help you deepen your understanding of content and prepare you to teach this investigation.

Investigation Overview

Students first explore their ideas about the Sun and compare their ideas to established knowledge about the Sun. Students learn more about the Sun's energy and its importance to the Earth as they undertake investigations at three different Solar Lab stations. **Digging Deeper** reviews information about the structure of the Sun. The reading also reviews the features found on and near the surface of the Sun.

Goals and Objectives

As a result of **Investigation 5** students will understand that the Sun is the major source of energy on which the Earth is dependent and that that energy consists of different wavelengths.

Science Content Objectives
Students will collect evidence that:
1. The Earth receives energy from the Sun.
2. Life on Earth is dependent upon energy from the Sun.
3. The Sun is a star — a mass of burning gases.

Inquiry Process Skills
Students will:
1. Compare their scientific explanations with established scientific explanations.
2. Use models to investigate science questions.
3. Collate information into a useful format.
4. Communicate observations and findings to others.

Teacher Commentary

Connections to Standards and Benchmarks

In **Investigation 5**, students examine their ideas about the Sun. They compare their ideas to reliable scientific sources. Students then work through mini-investigations in stations. The observations and investigations they complete will help them understand the National Science Education Standards and AAAS Benchmarks below:

NSES Links

- The Sun is the major source of energy for phenomena on the Earth's surface, such as growth of plants, winds, ocean currents, and the water cycle. Seasons result from variations in the amount of the Sun's energy hitting the surface, due to the tilt of the Earth's rotation on its axis and the length of the day.

- Gravity is the force that keeps planets in orbit around the Sun and governs the rest of the motion in the Solar System. Gravity alone holds us to the Earth's surface and explains the phenomena of the tides.

AAAS Links

- The Sun is a medium-sized star located near the edge of a disk-shaped galaxy of stars, part of which can be seen as a glowing band of light that spans the sky on a very clear night. The universe contains many billions of galaxies, and each galaxy contains many billions of stars. To the naked eye, even the closest of these galaxies is no more than a dim, fuzzy spot.

- The Sun is many thousands of times closer to the Earth than any other star. Light from the Sun takes a few minutes to reach the Earth, but light from the next nearest star takes a few years to arrive. The trip to that star would take the fastest rocket thousands of years. Some distant galaxies are so far away that their light takes several billion years to reach the Earth. People on Earth, therefore, see them as they were that long ago in the past.

- Because the Earth turns daily on an axis that is tilted relative to the plane of the Earth's yearly orbit around the Sun, sunlight falls more intensely on different parts of the Earth during the year. The difference in heating of the Earth's surface produces the planet's seasons and weather patterns.

Preparation and Materials Needed

Preparation

This investigation requires three 40-minute class periods to complete. **Day One:** Have the students address the **Key Question** and review the kinds of ideas they have and record them in their Journals. They can then review and complete **Part A** of the investigation. **Day Two:** Have students begin **Part B** of the three Solar Lab Stations of the investigation. **Day Three:** Have students rotate through the three Solar Lab Stations and then **Review and Reflect** on the whole investigation.

In **Investigation 5**, your students will share ideas about what they already know about the Sun and compare their ideas with the answers they find in the **Digging Deeper** section or in other reference material. Students will then explore the Sun's energy as they complete three activities at the Solar Lab Stations. **Station 1** allows students the opportunity to investigate the electromagnetic spectrum with the use of a prism. This will work best in an area that is slightly darkened. That will enable students to see the spectrum produced by the light and the prism more clearly.

At **Station 2**, students will investigate the impact the Sun's energy has on plants as they examine two plants that have been kept in different lighting conditions. This will require that you begin preparation about a week in advance of the lab. Use two of the same plants. (Swedish Ivy works well for this, as its leaves turn yellow fairly quickly when deprived of light.) Put one plant in a dark closet for a week and leave the second plant in the light. Make sure that both are watered regularly.

At **Station 3**, students will measure the heat from the Sun using thermometers. This investigation works best if there is direct access to sunlight. If your classroom does not have windows, or if it is a cloudy day, it is possible to simulate the experiment using heat lamps. Be sure students realize that this is just a simulation, however, and that actual sunlight may give slightly different results.

In **Part C** of the investigation, students will share their findings from the Solar Lab Stations and create a poster to share with the class.

This investigation requires three 40-minute class periods to complete, depending upon how you structure it. **Day One:** Have the students address the **Key Question** and review the kinds of ideas they have and record them in their Journals. They can then review and complete **Part A** of the investigation. **Day Two:** Have students begin **Part B** of the three Solar Lab Stations of the investigation. **Day Three:** Have students rotate through the three Solar Lab Stations and then complete **Part C** and **Review and Reflect** on the whole investigation.

Materials

Part B
Station 1
- triangular prism
- flashlight with a strong narrow beam
- colored pencils

Teacher Commentary

Station 2
- plant (provided by the teacher) that has been kept in a closet for a week
- matching plant that has been kept in the sunlight for a week

Station 3
- two alcohol thermometers
- window or access to the outdoors

Part C
- poster board
- colored pencils, markers

Investigating Earth in Space: Astronomy

Investigation 5: The Sun and Its Central Role in Our Solar System

Investigation 5:

The Sun and Its Central Role in Our Solar System

Key Question

Before you begin, first think about this key question.

What are the characteristics of the Sun and why is it so important to the Solar System?

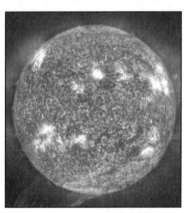

Think about what you know about the Sun. What is the Sun made of? Why is the Sun important to you?

Share your thinking with others in your group and with your class.

In this **Investigation**, you will become more familiar with three important facts about the Sun.

Investigate

Part A: What do you already know about the Sun?

1. You experience the Sun's energy every day. Take a minute and list in your journal all the ways you can think of that the Sun affects your life. (You might want to illustrate your list with pictures.) These sentence stems can help you in your thinking:

   a) The Sun is important because _____.

   b) If we didn't have the Sun, _____.

Teacher Commentary

Key Question

Write the **Key Question** on the board or on an overhead transparency. Encourage students to think about what they learned about the Sun and its relationship to the rest of the Solar System. Have them discuss this with members of their group. Tell students to record their ideas in a new journal entry.

Discuss student's ideas. Ask for a volunteer to record responses on the board or overhead projector transparency so that you can circulate among the students, encouraging them to copy the notes in an organized way.

Student Conceptions about the Sun

Students know that the Sun is extremely hot, is usually visible in the sky during the day, and that we see more of it in summer than in the winter. They will probably know that the Sun plays a major role in the seasons, but not necessarily how this happens. They may know that life depends on the Sun and that it provides energy. That said, don't be surprised if some students have very little understanding of what the Sun is like and why it is so important to the Solar System. Common misconceptions about the Sun include:
- The Sun rises exactly in the east and sets exactly in the west every day.
- The Sun disappears at night.
- We experience seasons because of the Earth's changing distance from the Sun (closer in the summer, farther in the winter).

Assessment Tool

Key Question-Evaluation Sheet
Use the evaluation sheet to help students understand and internalize basic expectations for the starting activity. The **Key Question-Evaluation Sheet** emphasizes that you want to see evidence of prior knowledge and that students should communicate their thinking clearly. You will not be likely to have time to apply this assessment every time students complete a starting activity; yet, in order to ensure that students value committing their initial conceptions to paper and taking the activity seriously, you should always remind them of the criteria. When time permits, use this evaluation sheet as a check on the quality of their work. As with any assessment tool used in *IES*, the instrument should be provided to students and discussed *before* they complete a task. This ensures that they have a clear understanding of your expectations for their work.

Answer for the Teacher Only

Our Sun is a low-mass star located in the Milky Way Galaxy. It is an enormous ball of glowing gas, containing many of the same materials found on Earth. These include hydrogen (71% by mass), helium (27%), and traces of the other elements (2%) such as calcium, sodium, magnesium, and iron. The fusion of hydrogen nuclei into helium nuclei is the reaction that produces the immense heat of the Sun (solar fusion). Even though the Sun looks small in the sky, it is actually very large. It seems small because it is 150 million kilometers (93 million miles) away. The Sun is roughly 1.4 million kilometers (900,000 miles) in diameter. In contrast, Earth's diameter is only 13 thousand kilometers (8000 miles).

The various wavelengths of the Sun's energy have different effects on our Earth, but some of the most familiar are the visible light spectrum, which enables us to see, and infrared radiation we experience as heat. Other types of energy from the Sun include ultraviolet radiation, x-rays, gamma rays, and cosmic rays.

The radiation from the Sun influenced the composition of the planets in the inner and outer regions of our Solar System. Because of its immense mass and associated gravitational pull, the Sun is the hub of the Solar System around which all other planets and objects orbit.

Investigate

Teaching Suggestions and Sample Answers

Part A: What do you already know about the Sun?
1. These questions should be used as a starting point for the students to list all the things that they know about the Sun. Be sure students are working together to share their ideas with the other members of their group.

Teacher Commentary

NOTES

INVESTIGATING EARTH IN SPACE: ASTRONOMY

Inquiry

Charts and Tables

Charts and tables organize a lot of information in a small amount of space. They can be useful because they allow you to focus on important points. In this investigation you will be using a table to organize what you know about the Sun and compare it to what scientists know.

2. Exchange your ideas about the Sun with other members of your group. Think about these questions:
   - What do we get from the Sun?
   - What place does the Sun have in our Solar System?
   - Where is the Sun in our galaxy?
   - What would we lose if we didn't have the Sun?
   - What reasons can you think of for some ancient peoples worshipping the Sun? Why was it so important to them? (Use your imagination with this question – why would the Sun be so important to early humans who knew very little about science?)

3. Make a three-column table in your journal to organize your thinking about the Sun.

   a) Write your answers to the questions in the table.

Revisit this table when you finish the Solar Lab and have read the **Digging Deeper** section. The Solar Lab will help you to understand more about the Sun, how it works, and why it is important in the Solar System. The Solar Lab is set up in stations, so that you and members of your group can work together to explore the Sun's energy and how it produces that energy.

Thinking about the Sun		
Sun Questions	My Answers	Scientists' Answers
What do we get from the Sun?		
What place does the Sun have in our Solar System?		
Where is the Sun in our galaxy?		
What would we lose if we didn't have the Sun?		
What reasons can you think of for some ancient peoples worshipping the Sun? Why was it so important to them?		
Other questions about the Sun.		

Teacher Commentary

1. Help students complete the sentences. They should do this individually at first, and then share what they have written with a partner and see if they can agree on a common response. Their combined effort can be now compared with another pair. Finally, hold a whole-class sharing of responses and see if everyone can agree on a final version. When you have this, post it for all to see. Revisit this at the end of this investigation to see if it still holds up.

2. Answers and responses will vary, and at times may be incorrect. That is all right. Encourage students to enter all their responses into the table. This will serve as a good learning tool later when they compare their answers with the actual scientific answers.

3. Introduce this table to your students and explain that it is a way to organize information into clear categories. Let them know that scientists frequently use tables in this way. Help students see that the sensible organization of information is crucial for a scientific investigation to succeed. Draw attention to the space at the bottom where students can add their own questions. Encourage them to make full use of this option.

Assessment Tools

Journal Entry-Evaluation Sheet
Use this sheet as a general guideline for assessing student journals, adapting it to your classroom if necessary. You should give the **Journal Entry-Evaluation Sheet** to students early in the module, discuss it with them, and use it to provide clear and prompt feedback.

Journal Entry-Checklist
Use this checklist for quickly checking the quality and completeness of journal entries. You can assign a value to each criterion, or assign a " + " or " − " for each category, which you can translate into points later.

Investigating Earth in Space: Astronomy

Investigation 5: The Sun and Its Central Role in Our Solar System

Part B: Solar Lab

Solar Lab Station 1: The Electromagnetic Spectrum

You have probably already learned a little about light in other science courses. Visible light is part of the electromagnetic (EM) spectrum – the energy we receive from our star, the Sun.

You can see from the diagram that the longer wavelengths are on the left side of the spectrum and that the wavelengths get shorter as you travel to the right side. You feel infrared radiation as heat. You may also be familiar with black light, which is actually ultraviolet radiation.

Materials Needed

For Station 1, your group will need:
- triangular prism
- flashlight with a strong narrow beam
- colored pencils

1. Shine the flashlight through the prism onto a dark surface and see for yourself what colors are in the visible light spectrum.

    a) Use colored pencils to draw in your journal what you see.

2. Use your drawing and the explanation above to answer the following questions:

    a) Which color of light seems to be bent the *most* by the prism?

    b) Which color of light seems to be bent the *least*?

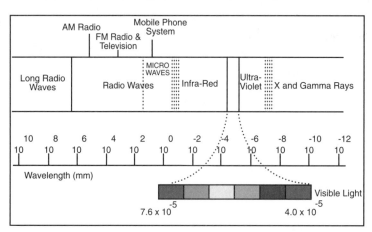

The electromagnetic spectrum shows how different wavelengths represent different types of energy.

Teacher Commentary

Part B: Solar Lab.
Solar Lab Station 1: The Electromagnetic Spectrum

Help students interpret the chart. Draw their attention to the visible light spectrum of colors. They may well be surprised at what a small portion of the spectrum this represents. The students should understand the visible light is only one form of electromagnetic radiation, as are radio waves and x-rays.

Make sure they understand how to use the chart to investigate the questions.

INVESTIGATING EARTH IN SPACE: ASTRONOMY

c) Of the colors in the visible light spectrum you can see, which has the longest wavelength?

3. Look again at the diagram of the EM spectrum.

   a) In your journal, make a list of ways that you think life on Earth would change if we didn't have this energy from the Sun.

Solar Lab Station 2: Plants and the Sun's Energy

At this station, you will see two plants that have the same type of soil and were watered the same amount every day.

1. Observe the two plants. One plant has been exposed to sunlight on a day to day basis, and the other plant has been kept in a closet for a week.

   a) What differences do you observe between the two plants? How might you explain this?

   b) What do you think would happen to the plant kept in the closet if it were never put in the Sun again? Why do you think that is so?

   c) What do you think the relationship is between the Sun and plants on Earth?

   d) How much do animals on the Earth depend on plants? What might happen to animal life on the planet without the Sun's energy?

2. As a group, design an experiment that you could do that would further test these ideas. Can you think of a way to test the impact of the sunlight on only the plant leaves?

   a) Write down your steps, the materials you would need, and your procedure for your teacher to review.

   b) Even if you do not conduct the experiment, what do you think you would be able to learn from this experience?

Materials Needed

For Station 2, your group will need:

- plant that has been kept in a closet for a week
- plant that has been kept in the sunlight

Teacher Commentary

1. Ask students to use colored pencils and draw the spectrum they find after using the prism.

2. a) Violet
   b) red
   c) red

3. Answers will vary but students will probably mention the fact that it would be colder and that plants would not grow as well.

Solar Lab Station 2: Plants and the Sun's Energy

Note that the plants need to have been prepared at least one week in advance. Another way to do this would be to place a cover (inverted plate or non see-through container) over a patch of grass for about a week. Students can then lift off the cover and observe the differences between the grass there and alongside.

1. a) Students should note that the plant that has been in the closet for a week has turned brown or yellow and is not as healthy looking as the other plant.
   b) Most students will probably guess that the plant, if kept in the closet, would eventually die.
   c) Plants on the Earth need the Sun to survive. Plants use sunlight to make the food they need to live.
   d) Most animal life on the planet depends on the Sun. Energy from the Sun provides heat and plants, as food energy, for the animals. Most animals, at the bottom of their food web, are dependent upon producers that make their food using light energy from the Sun.

Investigating Earth in Space: Astronomy

Investigation 5: The Sun and Its Central Role in Our Solar System

Solar Lab Station 3: Heat from the Sun

1. At this station, you will have the opportunity to investigate the Sun's infrared (heat) energy.

   a) Draw a table in your journal like the one below.

Temperature Readings (degrees Celsius)			
	0 minutes	5 minutes	10 minutes
Thermometer under paper			
Thermometer in direct sunlight			

2. There are two thermometers under a sheet of paper at this station.

   a) Lift the paper and record the temperatures of both thermometers.

   b) Place one thermometer back under the paper and put the second thermometer into direct sunlight. Wait five minutes, and then record the temperatures again.

Materials Needed

For Station 3, your group will need:
- two thermometers
- sheet of paper

Inquiry
Quantitative and Qualitative Observations

Observations dealing with numbers are called quantitative observations. An example of a quantitative observation is temperature recorded in degrees Celsius. Qualitative observations refer to the qualities of the object. Color is often recorded qualitatively as yellow or green, for example. Some observations can be made either qualitatively or quantitatively, depending on what tools are available and the level of accuracy needed. In this investigation you are making both qualitative and quantitative observations.

Teacher Commentary

2. Student ideas will vary. Lead them to come up with practical ideas which will allow them to make predictions and potentially test them.

Solar Lab Station 3: Heat from the Sun

It will help if the students are doing this on a bright sunny day. However, it should work even with a cloudy sky. If necessary, use a desk lamp to simulate the sunlight. Be careful that other variables do not come into play. For example, if the window has a heating or air conditioning outlet underneath it might affect the reliability of the experiment.

INVESTIGATING EARTH IN SPACE: ASTRONOMY

    c) Wait an additional five minutes, and then record the final temperatures.

3. Write the answers to the following questions in your journal:

    a) What can you conclude about the effect of the direct sunlight on the thermometer?

    b) Did the energy of the Sun have any effect at all on the thermometer under the paper? Why do you think that?

Part C: Sharing and Discussing Your Findings

1. When you finish all the stations in the Solar Lab, return to your table of questions about the Sun, from the beginning of this investigation.

    a) Change any answers that were incorrect or incomplete.

    b) Work with your group to come up with as complete a picture of the role of the Sun in the Solar System as possible.

2. As a group, create an informative, yet interesting and colorful, poster about the Sun. The poster should focus on how the Sun and the energy from the Sun has an impact on life on Earth as well as general facts about the Sun.

3. Share your posters with the other students in your class.

Materials Needed

For this part of the investigation your group will need:

- poster board and colored pencils

Teacher Commentary

3. a) Direct sunlight will increase the temperature reading on a thermometer faster than the thermometer that is not in the direct sunlight.
    b) Yes, the Sun did have an effect on the temperature of the thermometer under the paper. The Sun was not directly shining on it but it was warming up the air around the thermometer. Students will probably notice that the temperature shown on that thermometer will also increase, just not at the same rate.

Part C: Sharing and Discussing Your Findings

When all student groups have completed their work at the three Solar Lab Stations, have them revisit their data tables and complete the information. Next, have each group summarize their findings in a form that allows them to present it to others. Hold a class discussion in which each group in turn makes its report. Identify any questions or ideas that need to be clarified and work with your students to do this.

Investigating Earth in Space: Astronomy

Investigation 5: The Sun and Its Central Role in Our Solar System

 Digging Deeper

OUR SUN: A FAMILIAR OBJECT

The Sun is the probably the most prominent object in the sky because of its great size and the enormous amount of energy it releases. The Sun contains 99% of all the matter within the Solar System. It has 300,000 times more mass than the Earth and more than 1,295,000 times more volume. During the day, you can watch what appears to be the Sun's journey across the sky as the Earth receives the Sun's energy. This energy is a strong driving force in the Earth system. It heats the surface of the Earth and makes life possible. At night, the Sun's rays illuminate the Moon. Without the Sun, the Earth would be a dark frozen mass drifting through space.

The Reason For the Seasons

The tilt of the Earth's axis changes over the course of about 41,000 years between an angle of 22° and an angle of 25° as the Earth revolves around the Sun. For part of

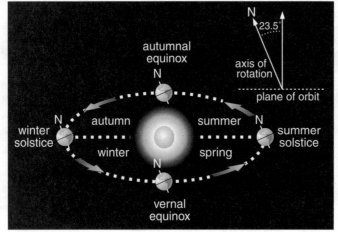

The tilt of the Earth on its axis and its rotation around the Sun cause the seasons.

Evidence for Ideas

As You Read...
Think about:

1. *How is the tilt of the Earth's axis related to the seasons?*
2. *What is the result of the spreading effect of the solar energy over the surface of the Earth?*
3. *Why is noon usually the hottest part of the day?*
4. *Why is the Sun important to our Solar System?*
5. *Where is our Sun in the universe?*
6. *What are some of the Sun's special features?*

Teacher Commentary

Digging Deeper

This section provides text, illustrations, and photographs that give students greater insight into the nature of the Sun. You may wish to assign the **As You Read** questions as homework to help the students focus on the major ideas in the text.

As You Read...
Think about:

1. The tilt of the Earth provides more direct sunlight for certain parts of the surface. This uneven heating of the surface of the Earth due to the tilt of the axis, helps play a role in the seasons on the Earth.

2. It causes uneven heating, making it warmer at the Equator and colder at the poles.

3. The Sun is high overhead and the energy is concentrated on the surface of the Earth.

4. The Sun provides heat and light to the Solar System. The Sun's gravity also holds the planets and other objects in the Solar System in their orbits.

5. The Sun is on the outside of one of the spiral arms of the Milky Way Galaxy.

6. Sunspots, solar flares, and prominences are all features found on the surface or directly above the surface of the Sun.

> **Assessment Opportunity**
>
> You may wish to rephrase selected questions from the **As You Read** section to use as quizzes. You could use multiple choice or true/false formats. This will provide assessment information about students' content understanding and can serve as a motivational tool to ensure that they complete and understand the reading assignment.

INVESTIGATING EARTH IN SPACE: ASTRONOMY

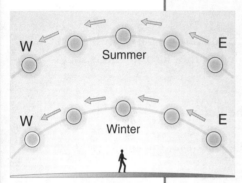

In the summer, the Sun is higher in the sky than in the winter.

the year (summer), the Northern Hemisphere tips towards the Sun. During that same time, the Southern Hemisphere is having winter, since it is tipped away from the Sun.

When either of the Northern or Southern Hemispheres is tipped towards the Sun, two things happen. First, the Sun is visible for more hours of the day, providing more heat energy. Second, the Sun is higher in the sky at noon and shines more directly on the Earth's surface than any other time.

Six months later, the Earth is halfway through its orbit and the same hemisphere is tilted away from the Sun. During this time, the Sun is lower in the sky and the days are shorter. Also, the Sun's energy is spread over the Earth's surface. These effects result in the cooler temperatures of winter. Each year, the cycle of the seasons repeats itself because of the regular and predictable orbit of the Earth around the Sun.

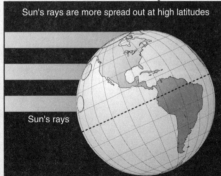

The Earth's tilt and curved surface cause the Sun's rays to be more spread out at high latitudes.

The Earth's curved surface affects how much of the Sun's (solar) energy the Earth receives. A greater angle between the Sun's rays and the Earth's surface causes energy to be spread over a larger area. When is the hottest part of day? It is usually around noon when the Sun is high in the sky and its energy is concentrated on the surface of the Earth.

The spreading effect of solar energy on the Earth's surface creates different temperatures at different points on the Earth. The direct concentration of solar energy on the Earth's Equator causes it to be much warmer than the poles. The spreading effect also causes the Northern and Southern Hemispheres to experience opposite seasons throughout the year.

Teacher Commentary

NOTES

Investigating Earth in Space: Astronomy

Investigation 5: The Sun and Its Central Role in Our Solar System

What is the Sun?

The Sun, the center of our Solar System, is an enormous ball of glowing gas. Most of the Sun is made of hydrogen, one of the simplest atoms in nature. It has a nucleus with one proton at its center and one electron orbiting around it. In the core of the Sun, these protons combine to create new nuclei of helium. This reaction is known as hydrogen fusion. It produces huge amounts of energy that cause the Sun to glow.

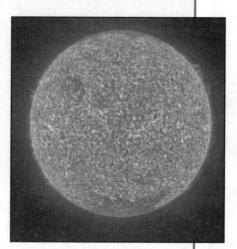

The reactions within the Sun produce a huge amount of energy that cause the Sun to glow.

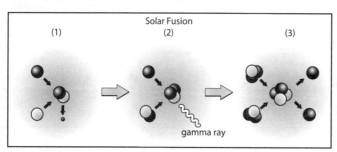

(1) Nuclei of two hydrogen atoms collide in a high-temperature environment. Particles and energy are released. A two-part atom is formed.
(2) Hydrogen nuclei collide with the two-part atom. A three-part helium atom is formed. More energy is released as well as gamma rays.
(3) Two three-part helium atoms collide to form a four-part helium atom. More energy is released as well as a pair of hydrogen atoms.

Structure of the Sun

Like the Earth, the Sun has a layered structure. The outer part of the Sun consists of an atmosphere known as the corona. This layer extends millions of kilometers into space. Temperatures in the corona can get as high as 1,000,000°C. Its gases are so hot that the Sun's gravity

Investigating Earth Systems
A 41

Teacher Commentary

> **About the Photo**
> The photo on page A41 of the student text shows the surface of the Sun. Many of the photos of the Sun that scientists gather today are made with telescopes on Earth. There are several major solar telescopes in the world, one of which is found at La Palma, in the Canary Islands off the coast of Africa.

INVESTIGATING EARTH IN SPACE: ASTRONOMY

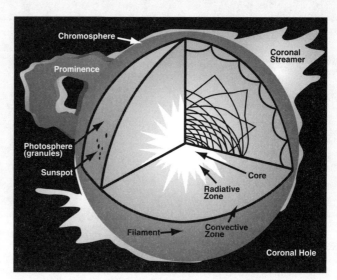

Temperatures vary in the Sun's different layers.

cannot hold them and they escape into space. Below the corona is the irregular surface of the Sun, called the photosphere. This is the visible surface of the Sun and the layer that produces light. Between these two layers is the chromosphere. The temperatures in this layer range from 6000°C to 20,000°C. This layer has a reddish appearance and extends about 2500 km above the photosphere.

Surface Features of the Sun

From the Earth, the Sun appears to have a smooth surface. This surface is actually not so smooth and contains a number of interesting features.

Sunspots are areas that have an irregular shape and are darker than other parts of the Sun. Sunspots are relatively cooler than the surrounding parts of the Sun, although they are still very hot. They develop in pairs and appear to travel across the surface of the Sun.

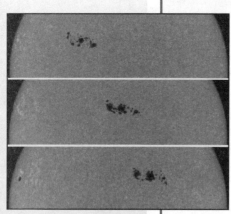

The dark irregular shapes on the surface of the Sun are sunspots.

Teacher Commentary

NOTES

Investigating Earth in Space: Astronomy

Investigation 5: The Sun and Its Central Role in Our Solar System

Their movement is actually due to the Sun spinning on its axis. Because the Sun is made up of gas, it rotates more quickly at the Equator (about 25 days) than at the poles (about 33 days). Sunspots at the poles take longer to travel around the Sun than at the Equator. The number of sunspots on the Sun increases and decreases during cycles that last about 22 years.

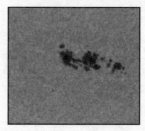
A close-up of a sunspot.

Solar flares are giant explosions of gas on the surface of the Sun. They occur near sunspots and erupt outwards with brilliant colors. During a solar flare, material is heated to millions of degrees Celsius in a matter of minutes. Then it is blasted off the surface. Many forms of energy are released during a solar flare. These include gamma rays and x-rays. Some of the most violent flares can produce enough radiation to be harmful to astronauts or damaging to satellites. Another

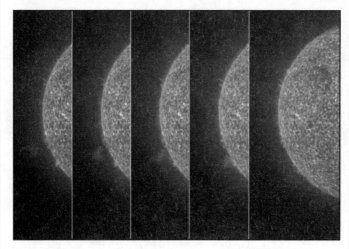
Prominences are huge arcing columns of gas that come from the Sun.

Teacher Commentary

NOTES

INVESTIGATING EARTH IN SPACE: ASTRONOMY

feature of the Sun's surface are huge rising columns of flaming gas called prominences. They are not quite as violent as solar flares and look like feathery red arches.

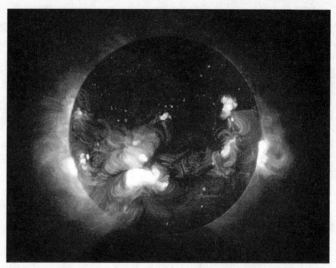

Hot gases in the corona of the Sun cause CMEs to occur.

Coronal mass ejections (CMEs) occur when gases in the corona are so hot that the Sun's gravity cannot hold them. The gases break free from the Sun and form a solar wind. In a single second, this wind can travel 400 km! CMEs take several hours to develop and create one of the largest features in the Sun's atmosphere. Each day, two or three CMEs can occur close to sunspot activity. The solar winds carry magnetic clouds with them. Some of the high-energy particles from these clouds reach the Earth. They can cause problems with communications equipment, including satellites.

Teacher Commentary

NOTES

Investigation 5: The Sun and Its Central Role in Our Solar System

Review and Reflect

Review

1. How does the tilt of the Earth produce the seasons?
2. Are the Sun's rays striking the Northern Hemisphere or the Southern Hemisphere more directly when it is summer in North America?
3. When is the hottest part of the day usually? Why?
4. What types of energy does the Sun produce?
5. What evidence do you have that plants need light?
6. What is the role of the Sun in our Solar System?

Reflect

7. How are you personally affected by the change in seasons?
8. Who would be affected more by the change in seasons, someone living close to the Equator or someone living in the far north? Explain.
9. How do humans use the different types of energy in the electromagnetic spectrum?
10. Do you think there would be life on Earth without the Sun? Why do you think that?

Thinking about the Earth System

11. The Sun is vital to the Earth System. How does the effect of the Sun on one Earth System affect other systems?

Thinking about Scientific Inquiry

12. In which parts of the investigation did you:
    a) Read for understanding?
    b) Record your own ideas?
    c) Use tools to make measurements?
    d) Make inferences from data?
    e) Share ideas with others?
    f) Find information from different sources?

Teacher Commentary

Review and Reflect

It is important that all students review carefully what they have done in this investigation, especially those things that have enhanced their understanding of the Sun and its importance in the Solar System.

Review

Allow your students ample time to pull all their evidence together and arrive at conclusions and explanations, and to make all the connections they can on the basis of their data. They should show a clear understanding of these ideas:

1. The tilt allows certain portions of the Earth's surface to receive more energy from the Sun.

2. The Northern Hemisphere.

3. Noon is usually the hottest part of the day because the Sun is directly overhead at that hour.

4. Light and heat energy.

5. Without light, plants die.

6. The role of the Sun in our Solar System is as the gravitational anchor holding the planets and other objects in orbit around it. The Sun also provides energy to the Solar System.

Reflect

7. Answers will vary.

8. Someone living to the north since they experience more seasonal fluctuations.

9. Humans use radio waves for AM and FM radio and television broadcasting and also for cell phones. They use infra-red waves for television remote control systems and heat lamps. They are also used by the military for night vision glasses and thermo-sensing devices. Ultra-violet light is used in sun-tanning salons, to encourage certain types of plant growth, by detectives to make fingerprints and other trace evidence visible, and for special lighting effects. X-rays are used in medical and industrial radiology to make internal parts visible. Gamma rays are best avoided completely by humans.

10. There could be no life on the planet without the Sun. All living things depend upon light energy transferred from the Sun through plants to the various food chains.

Thinking about the Earth System

11.
- In the *Hydrosphere* the Sun warms oceans and other surface water causing evaporation into the atmosphere.
- The Sun warms those parts of the *Atmosphere* near the Equator more and less so toward the North and South Poles. Through this, and because of Earth's rotation, air mass move causes weather patterns.
- Energy from the Sun enables life in the *Biosphere* and is the source for most food chains.
- The Sun affects the *Geosphere* directly in heating arid areas making them prone to wind erosion.
- The Sun affects the *Geosphere* indirectly by its effects on the other systems which in turn affect the *Geosphere* (erosion and deposition caused by water, the effect of the *Biosphere* on the *Geosphere*, etc.)

Teacher Commentary

NOTES

Teacher Review

Use this section to reflect on and review the investigation. Keep in mind that your notes here are likely to be especially helpful when you teach this investigation again. Questions listed here are examples only.

Student Achievement

What evidence do you have that all students have met the science content objectives?

Are there any students who need more help in reaching these objectives? If so, how can you provide this?

What evidence do you have that all students have demonstrated their understanding of the inquiry processes?

Which of these inquiry objectives do your students need to improve upon in future investigations?

What evidence do the journal entries contain about what your students learned from this investigation?

Planning

How well did this investigation fit into your class time?

What changes can you make to improve your planning next time?

Guiding and Facilitating Learning

How well did you focus and support inquiry while interacting with students?

What changes can you make to improve classroom management for the next investigation or the next time you teach this investigation?

Teacher Review

How successful were you in encouraging all students to participate fully in science learning? _____

_____

How did you encourage and model the skills values, and attitudes of scientific inquiry? _____

_____

How did you nurture collaboration among students? _____

_____

Materials and Resources

What challenges did you encounter obtaining or using materials and/or resources needed for the activity? _____

_____

What changes can you make to better obtain and better manage materials and resources next time? _____

_____

Student Evaluation

Describe how you evaluated student progress. What worked well? What needs to be improved? _____

_____

How will you adapt your evaluation methods for next time? _____

_____

Describe how you guided students in self-assessment. _____

_____

Self Evaluation

How would you rate your teaching of this investigation? _____

_____

What advice would you give to a colleague who is planning to teach this investigation? _____

_____

NOTES

Teacher Commentary

INVESTIGATION 6: THE PLANETARY COUNCIL

Background Information

Mercury

The planet Mercury is very difficult to study from the Earth because it is always so close to the Sun. It is never more than 28 degrees from the Sun in our sky. It is the second smallest planet. It was believed to have been the smallest until scientists discovered that Pluto is actually much smaller than it was originally thought. Mercury is the fastest-moving planet in its orbit around the Sun. It takes just 88 Earth days for Mercury to travel around the Sun. And also the fastest in its orbit since it is the innermost planet. In fact, the name Mercury derives from its speed in moving around its orbit.

We began to learn more about Mercury with radar imaging from the Earth in the 1960s, and obtained most of what we know about the planet from the Mariner 10 space probe was placed into a complicated orbit involving Venus and Mercury and which passed close to Mercury and sent back information three times in the period 1974–1976.

Mercury rotates on its axis once every 59 Earth days. This slow rate, along with the proximity to the Sun, allows surface temperatures of up to 400°C in the daytime. At night, the lack of an atmosphere allows most of the heat to escape, bringing temperatures to as low as –200°C.

Mercury has no atmosphere because its gravitational pull is very weak. Plus, the high temperatures make particles move very quickly.

Venus

Venus is the only planet to rotate from east to west. Images have shown that the surface of Venus is in many ways similar to the surface of Earth. Lava flows, faults, basins, mountains, and volcanoes dot the surface of Venus. It is believed that the oldest crust on Venus is 800 million years old. The atmosphere of Venus is very hostile. It is made mostly of carbon dioxide with a small percentage of nitrogen. Venus is surrounded by yellow clouds of sulfuric acid. It is estimated that atmospheric pressure on the surface of Venus is about 90 times greater than it is on Earth. The surface of Venus gets very hot, due to the greenhouse effect. Surface temperatures of 482°C can be reached.

Mars

Mars is about half the size of Earth. Its gravitational pull is less than that on Earth and it has a very weak magnetic field. Mars does have seasons, much like on Earth. It is tilted at almost the same angle and direction as the Earth is on its axis. Each season on Mars is twice as long as it is on Earth, however. The atmosphere of Mars is about 95% carbon dioxide and roughly 5% nitrogen and argon. Atmospheric pressure on Mars is about 150 times less than it is on Earth.

Mars has polar ice caps made of frozen carbon dioxide and small amounts of frozen water. Mars has extinct volcanoes and a substantial canyon system.

Recently, probes have landed on the surface of Mars and moved around. These have sent back pictures and data on the composition of the solid and rocks.

Jupiter

Jupiter rotates faster than any other planet — once every 10 Earth hours. It is the largest planet in the entire Solar System and if all the other planets were combined, the mass of those would only be about half of that of Jupiter.

Jupiter has a gaseous surface with areas of rising and sinking gas. The wind on Jupiter is very swift.

One of the most striking features is that of the Great Red Spot. This is a massive storm which rises about 8 kilometers above the clouds. Jupiter also has a very strong magnetic field. Colorful light displays, like the aurora borealis on Earth, are frequent on Jupiter. Jupiter has many moons, at least 16 of them.

Saturn

Saturn is also a gas planet, with bands of swirling gas on the surface. The density of the planet is quite low. The winds at the equator of Saturn move at speeds upward of 1800 km/hour. Saturn has some sources of internal heat and a weak magnetic field.

Saturn's most recognizable features are its rings, which are composed of rock and ice fragments.

Uranus

Uranus was discovered in 1751 because of its distance from the Earth.

Uranus is tipped almost completely over, rotating on its axis sideways. Scientists believe that the planet became tipped as a result of a collision with an object about the size of Earth during the formation of the Solar System.

Uranus also has a magnetic field. Most planets have a magnetic field that is basically in the same direction as its axis. Uranus, however, is different. Its magnetic field is about 60° off of its axis. This causes the magnetic field of Uranus to trace a spiral pattern in the solar wind as the planet rotates.

The average temperature on Uranus is −200°C. The surface temperature on Uranus is basically uniform over the entire planet. It seems as if there is some atmospheric cause of these phenomena.

Neptune

Neptune was discovered in 1846. Wind speeds on Neptune have been measured up to 2200 km/hr. Neptune also gives off 2.7 x more energy than it receives from the Sun.

Pluto

Pluto was discovered in 1930. Surface temperatures on Pluto are estimated to be at −220°C. At this temperature, gases such as methane and helium are frozen. Pluto is made of water, ice, and rocks. Pluto and Neptune cross each other in their orbits, at times making Pluto closer to the Sun than Neptune.

More Information…on the Web

Visit the *Investigating Earth Systems* web site www.agiweb.org/ies for links to a variety of web sites that will help you deepen your understanding of content and prepare you to teach this investigation.

Teacher Commentary

Investigation Overview
In **Investigation 6**, student groups will specialize in learning about a particular planet so that they can defend why their planet should receive federal funding to be studied. **Digging Deeper** examines other bodies found in the Solar System, such as meteoroids, asteroids, and comets. The reading also discusses the nebular theory and the arrangement of the Solar System.

Goals and Objectives
As a result of **Investigation 6**, students will understand the features of their own planet as well as those of the remaining planets in the Solar System.

Science Content Objectives
Students will collect information about:
1. A planet in which their group specializes.
2. The remaining planets in the Solar System.
3. The ways in which planets have been studied to date.

Inquiry Process Skills
Students will:
1. Locate and review relevant information resources.
2. Identify key scientific features relating to an object.
3. Communicate observations and findings to others.

Connections to Standards and Benchmarks

In **Investigation 6**, students apply what they have already learned to a chosen planet in the Solar System. The observations and investigations they complete will help them understand the National Science Education Standards and AAAS Benchmarks below:

NSES Links

- The Earth is the third planet from the Sun in a system that includes the Moon, the Sun, eight other planets and their moons, and smaller objects, such as asteroids and comets. The Sun, an average star, is the central and largest body in the Solar System.

- Most objects in the Solar System are in regular and predictable motion. Those motions explain such phenomena as the day, the year, phases of the Moon, and eclipses.

- Gravity is the force that keeps planets in orbit around the Sun and governs the rest of the motion in the Solar System.

AAAS Links

- Nine planets of very different size, composition, and surface features move around the Sun in nearly circular orbits. Some planets have a great variety of moons and even flat rings of rock and ice particles orbiting around them. Some of these planets and moons show evidence of geologic activity. The Earth is orbited by one Moon, many artificial satellites, and debris.

Teacher Commentary

Preparation and Materials Needed

Preparation
Students will need to have access to a variety of resources for this investigation. One of the best resources is the NASA web site: www.nasa.gov, but they can also use books, journals, CD-ROMs, and videos. You will need to allot time for students to prepare a presentation on what they have learned about their planets and how it has been studied in the past. They will also need presentation materials, and, possibly, computer access.

This investigation requires six 40-minute class periods to complete, depending upon how you structure it. **Day One:** Have the students address the **Key Question** and review the kinds of ideas they have and record them in their Journals. They can then review the whole investigation and make plans on how to proceed. **Day Two:** Students will perform the activity in which they model the movement of planets within an orbit. **Day Three:** Have students begin their investigation into their chosen planets. **Day Four and Day Five:** Have students complete their planet investigations and prepare their presentations. **Day Six:** Arrange for each group of students to make their presentations and **Review and Reflect** on the investigation as a whole.

Suggested Materials
- resource materials on the planets (NASA web site, books, journals, videos, CD-ROMs, etc.)
- access to the Internet (highly recommended)
- presentation materials (overheads and markers, PowerPoint™ software, computer, computer projector, poster board and markers, photographs, VCR and monitor)

INVESTIGATING EARTH IN SPACE: ASTRONOMY

Investigation 6:
The Planetary Council

Key Question:
Before you begin, first think about this key question.

How are the planets in our Solar System the same and how are they different?

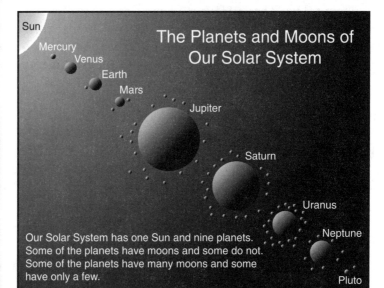

The Planets and Moons of Our Solar System

Our Solar System has one Sun and nine planets. Some of the planets have moons and some do not. Some of the planets have many moons and some have only a few.

Think about what you have learned so far about the planets in the Solar System. What are the names of the planets? Are they different sizes? Which ones are large and which ones are small?

Share your thinking with others in your group and in the class.

Teacher Commentary

Key Question

Refer students to **Investigation 4**, in which they made a model of the Solar System that included each of the planets. Because of this, they should not have much difficulty in finding some reasonable answers to this **Key Question**. However, there are many more similarities and differences to be found than they will have had the opportunity to discover so far. Use this question both to establish what they know about the planets and what they do not.

Write the **Key Question** on the chalkboard or on an overhead transparency. Have the students record their answers in their journals. Tell them to think about and answer the question individually. Ask them to write as much as they know and to provide as much detail as possible in their responses. Emphasize that they should include the current date and the prompt (the question itself, a meaningful heading, etc) in their journal entries.

> **Assessment Tools**
> Journal Entry-Evaluation Sheet
> Remember to draw students' attention to this tool. (**Blackline Master** is included in the Appendix.)

Student Conceptions about the Solar System

By now, students should be in a good position to complete this investigation. Many of their early ideas about Earth and Space will have been clarified and extended. Nevertheless, be alert to student understandings that need further teaching and help them toward this goal.

Answer for the Teacher Only

The planets in our Solar System are similar in that they formed from the same nebular source and all are in orbit around the Sun and receive energy from it. They are different in a variety of ways: distance from the Sun and from each other; size; structure; length of orbit around the Sun; number of moons; surface temperatures; presence or absence of rings; appearance as viewed from the Earth, etc.

Investigating Earth in Space: Astronomy

Investigation 6: The Planetary Council

Investigate

In **Investigation 4,** you started to think about the planets in our Solar System. You studied their location in relation to one another and their different sizes. In this investigation, you will learn how the planets are alike and how they are different.

1. In this investigation, you will be role-playing researchers in a space program. Each group in your class will represent a different planet in our Solar System (excluding the Earth). There is only a limited amount of funding to explore the different planets. Each group wants as much research funding as possible for exploring "its planet." For this to happen, you will need to prove to the Planetary Council that your planet should be studied in depth. You will need to make a professional presentation that covers the following points:

    • the features of your planet (size, atmosphere, number of moons, rings, etc.)

    • what evidence there is for how your planet formed

Materials Needed

For this investigation, your group will need:

• resource materials on the planets

• access to the Internet (optional)

• presentation materials (overheads and markers, PowerPoint™ software, computer, computer projector, poster board and markers, photographs, VCR and monitor)

Investigating Earth Systems

172 Investigating Earth Systems

Teacher Commentary

Investigate

Teaching Suggestions and Sample Answers

Planetary Council

This investigation uses role-play as a method of motivating students to find out about the properties and characteristics of the eight remaining planets: Mercury, Venus, Mars, Jupiter, Saturn, Uranus, Neptune, and Pluto. This should be quite exciting, and it is important that the role-play be maintained throughout. Be creative in the way you work with students in this respect. It may be helpful for you to play a role yourself such as Chair of the Planetary Council, Chair of the Funding Council, Director of Planetary Research, Expedition Controller, or even the Administration's Vice President. At the end, the Funding Council decides which planet will receive the funding. You may want to invite other teachers or members of the community to sit on the Council.

1. Sharing responsibility is an important aspect of this investigation and mirrors what scientists frequently do. Advise groups to "play to their strengths" by seeing the mix of talents and interests they have and using these for the collective good.

INVESTIGATING EARTH IN SPACE: ASTRONOMY

- where the planet is in the Solar System in relation to the Sun and the other planets
- what your planet looks like (use photographs, drawings, video, etc.)
- what interesting questions your research team wants to find out about your planet
- what other objects in, or passing through, the Solar System might have an effect on your planet
- what technology you think could be used to study and/or explore your planet.

Teacher Commentary

Teaching Tip

Group Participation Evaluation Forms I and II

One of the challenges to assessing student who work in collaborative teams is assessing group participations. Students need to know that each group member must pull his or her weight. As a component of a complete assessment system, especially in a collaborative learning environment, it is often helpful to engage students in a self-assessment of their participation in the group. Knowing that their contributions to the group will be evaluated provides an additional motivational tool to keep students constructively engaged.

Group Participation Evaluation Forms I and II provides students with an opportunity to assess group participation. In no case should the results of this evaluation be used as the sole source of assessment data. Rather, it is better to assign a weight to the results of this evaluation and factor it in with other sources of assessment data. If you have not done this before, you may be surprised to find how honestly students will critique their own work, often more intensely than you might do.

Investigation 6: The Planetary Council

2. Once groups have chosen their planets, you will need to decide on research assignments for each group member. Your teacher will have a list of web sites for you to use, and you can also use your school's media center and classroom resources.

3. Plan your presentation. Remember that you want to make the strongest case possible for your planet. You might want to use a PowerPoint™ program to present your information. You might also want to use posters, photos, videos, or overheads. As you research your planet, find out what questions scientists have been asking about it over the years. You will also need to know the latest discoveries that scientists have made about your planet.

4. When you finish your research, outline the important points you need to make in your presentation. Divide up the work fairly, put your presentation together, and rehearse it.

5. Your teacher will arrange a time for the presentations to the Planetary Council. You will need to do a convincing job, so look and act like professional scientists. Be sure to take notes on what the other groups present. That way you can identify what features are the same from planet to planet and what features are different.

6. After each presentation, answer the following questions in your journal:

   a) What is the size of the planet?

   b) How far is it from the Sun?

   c) What is its atmosphere like?

   d) How many moons does the planet have?

   e) What are the unique features of the planet?

   f) What does the planet look like?

   g) What kind of geologic action (if any) occurs on the planet?

7. When all groups have presented, the Planetary Council will decide which groups will receive the major funding.

Inquiry

Ways of Packaging Information

Scientists are often asked to provide information to the public or to make presentations. In doing so, they need to consider both the information they want to communicate and the person or groups that will be using the information. Then they must decide on the best method of packaging and delivering that information. These are decisions you need to make in this investigation.

Teacher Commentary

2. Help students to see that deciding upon the presentation method in advance ensures that the kind of information gathered will fit. Using PowerPoint (PPT) is strongly recommended because it is easy to use, can give very professional-looking results, and has many other features that can bring a presentation to life. Scientists regularly use PPT in their presentations to colleagues and funding agencies. Having students proficient in PPT and other useful computer software programs is a legitimate part of scientific study. The other presentational forms given here may also be used creatively and students may well come up with their own alternatives. Be alert to this aspect of the investigation from the start and encourage diversity and creativity.

3. Fixing the presentational content is an important step. Students may not have much experience with this and may need some help. Ask for the highest standard of presentation, whatever form it takes. Accepting less than polished work gives a message that "anything will do" and, in the end, is counter-productive.

4. Make sure that you set aside ample time for your students to make their presentations. Keep in mind that the presentations are likely to raise questions and discussion among the students. You may also wish to budget time for an "end of investigation" discussion.

5. The Funding Council must be able to defend its decision. Make sure that the reasons for choosing one planet rather than another are clearly stated. The Funding Council should feel that it has quite a bit of latitude in making its decision. For example, if several groups of students make excellent and convincing presentations, the Council may wish to award partial funding to several groups.

Investigating Earth in Space: Astronomy

INVESTIGATING EARTH IN SPACE: ASTRONOMY

As You Read...
Think about:
1. What is the difference between meteors, meteoroids, and meteorites?
2. What happens when a meteor strikes the Earth?
3. What are asteroids? Can an asteroid strike the Earth?
4. What are comets?
5. What is the relationship between a nebula and the objects in our Solar System?

Digging Deeper

METEOROIDS

Have you ever looked into the night sky and seen the bright streak of a shooting star? If you have, what you have really observed is a small meteoroid entering the Earth's atmosphere.

A shooting star streaking through the night sky

Meteoroids are small, rocky bodies that revolve around the Sun. If the Earth's orbit crosses the orbit of a meteoroid, the meteoroid may enter into the Earth's atmosphere. When this happens, the meteoroid starts to burn up and creates a streak of light. During its journey through Earth's atmosphere, the meteoroid is known as a meteor. If the meteoroid does not completely burn up and hits the Earth's ground, it is called a meteorite. Each year, the Earth gains about ninety million kilograms of matter from meteorites. Most are small specks, but some are quite large. The largest meteorite ever found was in Africa and weighed more than 54,000 kg!

Meteoroids enter the atmosphere at speeds of several thousand meters per second. The friction of the atmosphere slows down most meteoroids. Only meteoroids larger than a few hundred tons make craters when they strike the Earth. One example of a meteorite crater is in Arizona. This crater formed about 50,000 years ago and is 1200 m in diameter and 50 m deep.

Teacher Commentary

Digging Deeper

This section provides text, illustrations, and photographs that give students greater insight into the nature of the other objects in the atmosphere, including meteors, asteroids, and comets. Again, you could divide this **Digging Deeper** into sections and have different groups of students specialize in one section. However, it will be important to ensure that the gathered information is shared by all so that everyone can understand the full picture. Be sure to allow time for a full discussion after students read the **Digging Deeper** section. Be alert for any student misunderstandings and find ways to clarify the concepts involved. You may wish to assign the **As You Read** questions as homework to help students focus on the major ideas in the text.

As You Read...
Think about:

1. Meteors are meteoroids that enter the Earth's atmosphere. Meteoroids are small, rocky bodies that revolve around the Sun. Meteorites are meteoroids that enter the Earth's atmosphere and do not burn up completely and therefore crash to the surface of the Earth.

2. When a meteor hits the surface of the Earth, it creates a crater and kicks up a large amount of dust, or water, depending on where it lands.

3. Asteroids are bodies of metallic and rocky material that travel in orbits around the Sun, primarily between Mars and Jupiter. Asteroids have been known to hit the Earth.

4. Comets are small bodies of ice, rock and dust that are loosely packed together and that travel around the Sun on an orbit.

5. It is believed that our Solar System and all the material within the Solar System originated from a giant nebula, or cloud of dust and gas.

Investigation 6: The Planetary Council

ASTEROIDS

Asteroids are bodies of metallic and rocky material, sometimes called minor planets. They orbit the Sun like meteoroids, but are much larger in size. Most asteroids are located in a wide region of the Solar System called the Main Asteroid Belt. This belt of asteroids is like a giant doughnut-shaped ring between the orbits of Mars and Jupiter. During the early development of the Solar System, the strong gravity of Jupiter kept the asteroids in the Main Asteroid Belt from forming into a planet.

The Barringer meteorite impact crater in Arizona.

Asteroids that come close to our planet are known as near-Earth asteroids. The largest near-Earth asteroid is called 1036 Ganymede, and is 41 km in diameter. Scientists think that near-Earth asteroids are produced by the collision of asteroids within the Main Asteroid Belt. At least a thousand asteroids larger than 1 km in diameter have orbits that cross the orbit of the Earth. Sometimes these near-Earth asteroids collide with the Earth. About 65 million years ago, an asteroid 10 km across struck the Earth. The after-effects of this collision may have been responsible for the extinction of the dinosaurs.

A close-up of an asteroid.

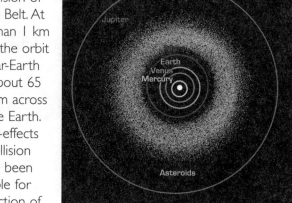

The Main Asteroid Belt is located between Mars and Jupiter.

Teacher Commentary

NOTES

INVESTIGATING EARTH IN SPACE: ASTRONOMY

Ice turns into a gas to form the streaming tail of a comet.

COMETS

Most comets can only be seen with a telescope but some pass close enough to Earth to be observed with the naked eye. A very bright comet can be seen in the night sky for days, weeks, or even months. A comet is a small body of ice, rock and dust that is loosely packed together. It can be thought of as an enormous snowball made of frozen gases that contains very little solid material. Comets, like planets, also orbit the Sun.

When a comet comes close enough to the Sun, the solar energy turns some of the ice into a gas. This vapor can be seen streaming behind the comet like a tail. Although very low in mass, comets are one of the largest members of the Solar System. The frozen part of a comet is only a few tens of kilometers at most. A comet's head can be as large as 100,000 km across. Its tail can be tens of millions of kilometers long.

One of the most famous of Earth's visiting comets is called Halley's comet, which was last seen in 1986. It passes by Earth every 76 years and will next appear in 2062.

NEBULAR THEORY

Most scientists agree that the objects in the Solar System formed about 4.6 billion years ago from a giant cloud of swirling gas and dust. This cloud is called a nebula, and its matter was probably thrown off from other stars in our region of the galaxy. Gravity caused the gases and dust to be drawn together, making the giant cloud fall inwards. As it collapsed, it got flatter,

Teacher Commentary

This section on Nebular Theory will help students understand how the planets and other objects in the Solar System formed long ago. This has relevance for all their chosen planets and will help them in constructing the case for each.

Investigation 6: The Planetary Council

 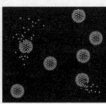 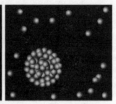

Particles within the nebula cloud were pulled together by gravity.

began to rotate and took the shape of a disk. The collapsing matter in the center of the disk eventually formed the Sun.

In the hot, inner part of the young Solar System, rock and metals with high freezing-point temperature remained solid. Here, the terrestrial planets formed as dense rocky worlds. The terrestrial planets include Mercury, Venus, Earth, and Mars. Farther from the Sun, the temperatures were lower. At this point, gas and icy particles (bits of matter) formed different types of planets. As the mass and gravity of these planets increased, they started to attract more particles. They

The Sun formed from a giant cloud of swirling gas and dust.

grew into the gigantic outer planets we call the gas giants. The gas giants include Jupiter, Saturn, Uranus, and Neptune.

Powerful telescopes have provided scientists with images of other stars and planets forming in the depths of giant clouds and rotating disks of gas and dust. Scientists have used the images as evidence of nebular theory.

Chandra is an x-ray telescope sent into orbit around the Earth in 1999.

Teacher Commentary

NOTES

INVESTIGATING EARTH IN SPACE: ASTRONOMY

Review and Reflect

Review

1. What did you learn about your planet that you did not know before?
2. What important and interesting questions have scientists asked about your planet?

Reflect

3. What explanation do you have for the similarities between the planets?
4. How are the planets different? Make a list of those differences as well.
5. How can you explain the differences between the planets?

Thinking about the Earth System

6. How are meteoroids and asteroids connected to the Earth System?
7. How does the Nebular Theory help you understand the structure of the Earth?

Thinking about Scientific Inquiry

8. In which parts of the investigation did you:
    a) Ask your own questions?
    b) Record your own ideas?
    c) Revise your ideas?
    d) Use your imagination?
    e) Share ideas with others?
    f) Find information from different sources?
    g) Pull your information together to make a presentation?

> Evidence for Ideas

Teacher Commentary

Review and Reflect

Review

The questions given in the Student Edition will drive this review. Here they focus on the particular planet each group of students researched. In answering the questions, your students will reveal what they have learned and also what they may have missed. Follow their responses carefully, noting any points of confusion or uncertainty. Where questions emerge, help to clarify ideas and concepts for your students.

1. In general, the larger the orbit, the longer the period of revolution. Answers will vary. Be sure that students' answers are well presented and correct.
2. Answers will vary. Be sure that students' answers are well presented and correct.

Reflect

Here students reflect on what they have learned about all the planets. The information will have come from the presentations made by other groups. Since this is a variable source, check that students have gained the targeted understandings. Provide clarification where needed.

3. The planets were formed at roughly the same time.
4. The inner planets are small and rocky. The outer planets, with the exception of Pluto, are large and composed mainly of gas. Students may have specific differences.
5. Many of the differences are due to the location of the planet with respect to the Sun.

Thinking about the Earth System

6. • Meteoroids orbiting the Sun can fall in the path of Earth's orbit. When they enter Earth's atmosphere they burn up, causing a streak of light which we call "shooting stars."
   • Some meteoroids manage to pass through Earth's atmosphere and become meteorites that crash into its surface, making craters. This can have an impact on all of Earth's systems within the vicinity and beyond, depending on the size of the meteorite and where it lands.
   • Asteroids, bodies of metallic and rocky material, are much larger than meteoroids. Earth's history shows that asteroids can also enter Earth's atmosphere and crash on its surface, but these are rare events. The last known asteroid strike was about 65 million years ago and had devastating effects on many living things in its vicinity and elsewhere on the planet.
7. Nebular theory offers a scientific explanation of how the Solar System and its planets, including Earth, were formed. This, in turn, offers an explanation of how galaxies are born and behave over time.

Thinking about Scientific Inquiry

For this investigation, it might be useful to have students first respond to questions individually. They can write these in their journals or on scratch paper. Next, hold a class discussion about each of the items, inviting students to contribute what they have written and the ideas behind them. Have the **Inquiry Processes** page open in their books for checking, or use the **Blackline Master Astronomy 1.1** in the Appendix. Once you finish the class discussions, allow students to revise their initial responses.

Teacher Review

Use this section to reflect on and review the investigation. Keep in mind that your notes here are likely to be especially helpful when you teach this investigation again. Questions listed here are examples only.

Student Achievement

What evidence do you have that all students have met the science content objectives?

_____

Are there any students who need more help in reaching these objectives? If so, how can you provide this? _____

_____

What evidence do you have that all students have demonstrated their understanding of the inquiry processes? _____

_____

Which of these inquiry objectives do your students need to improve upon in future investigations? _____

_____

What evidence do the journal entries contain about what your students learned from this investigation? _____

_____

Planning

How well did this investigation fit into your class time? _____

_____

What changes can you make to improve your planning next time? _____

_____

Guiding and Facilitating Learning

How well did you focus and support inquiry while interacting with students?

_____

What changes can you make to improve classroom management for the next investigation or the next time you teach this investigation? _____

_____

Teacher Review

How successful were you in encouraging all students to participate fully in science learning? _____

_____

How did you encourage and model the skills values, and attitudes of scientific inquiry? _____

_____

How did you nurture collaboration among students? _____

_____

Materials and Resources

What challenges did you encounter obtaining or using materials and/or resources needed for the activity? _____

_____

What changes can you make to better obtain and better manage materials and resources next time? _____

_____

Student Evaluation

Describe how you evaluated student progress. What worked well? What needs to be improved? _____

_____

How will you adapt your evaluation methods for next time? _____

_____

Describe how you guided students in self-assessment. _____

_____

Self Evaluation

How would you rate your teaching of this investigation? _____

_____

What advice would you give to a colleague who is planning to teach this investigation? _____

_____

NOTES

Teacher Commentary

INVESTIGATION 7: WHAT IS BEYOND OUR SOLAR SYSTEM?
Background Information

The Milky Way is the galaxy which is the home of our Solar System together with at least 200 billion other stars, although more recent estimates have given numbers around 400 billion. It includes those stars and their planets as well as thousands of clusters and nebulae. All the objects in the Milky Way Galaxy orbit a common center of mass, called the Galactic Center.

The Milky Way is a spiral galaxy whose spiral arms of our Milky Way contain interstellar matter, diffuse nebulae, and young stars and open star clusters emerging from this matter. Our Solar System is situated within the outer regions of this galaxy about 28,000 light-years from the Galactic Center. This makes the rest of the Milky Way visible from Earth. The center of the Milky Way Galaxy lies in the direction of the constellation Sagittarius, but very close to the border of both neighbor constellations Scorpius and Ophiuchus.

Stars are hot bodies of glowing gas that start their life in nebulae. They vary in size, mass and temperature, diameters ranging from 450 x smaller to over 1000 x larger than that of the Sun. Surface temperatures of stars can range from 3000° C to over 50,000° C. The color of a star is determined by its temperature, the hottest stars are blue and the coolest stars are red. The Sun has a surface temperature of 5500° C and its color appears yellow.

The energy produced by the star is by nuclear fusion in the stars core. The brightness is measured in magnitude; the brighter the star the lower the magnitude goes down. There are two ways to measure the brightness of a star: apparent magnitude is the brightness seen from Earth, and absolute magnitude which is the brightness of a star seen from a standard distance of 10 parsecs (32.6 light-years).

Stars begin their life cycles as high density nebula which condense into huge globules of gas and dust. These nebulas start to contract under its own gravity. Then a region of condensing matter will begin to heat up and start to glow forming a protostar. If a protostar contains enough matter the central temperature reaches 15 million degrees centigrade. This is the temperature at which hydrogen fuses to form helium in a nuclear reaction.

The star begins to release energy, stopping it from contracting even more and causes it to shine. It is now a Main-Sequence Star. Most stars remain main-sequence stars for about 10 billion years until all of the hydrogen has fused to form helium. The helium core starts to contract further and reactions begin to occur in a shell around the core. The core is hot enough for the helium to fuse to form carbon. The outer layers begin to expand, cool and shine less brightly. The expanding star is now called a Red Giant. Eventually, the helium core runs out, and the outer layers drift away from the core as a gaseous shell. The gas that surrounds the core is called a Planetary Nebula. The remaining core, which represents about 80% of the original star, is now in its final stages. The core becomes a White Dwarf. The star eventually cools and dims. When it stops shining, the now dead star is called a Black Dwarf.

If a star is particularly massive to begin with, the core will collapse in less than a second. This creates an explosion called a supernova. Amazingly, sometimes the core survives the explosion. If the surviving core is between 1.5–3 solar masses it contracts to become a tiny, very dense Neutron Star. If the core is much greater than 3 solar masses, the core contracts to become a Black Hole.

More Information…on the Web
Visit the *Investigating Earth Systems* web site www.agiweb.org/ies for links to a variety of other web sites that will help you deepen your understanding of content and prepare you to teach this module.

Teacher Commentary

Investigation Overview
Students conduct a series of investigations to learn about the appearance of stars in the night sky, constellations, galaxies, nebulae, and the origin of the universe. **Digging Deeper** reviews information about the different types of galaxies, with particular attention paid to our galaxy, the Milky Way Galaxy. The reading also reviews the different types of nebulae, the theories about the origin of the universe, and the life cycle of the stars.

Goals and Objectives
As a result of **Investigation 7**, students will understand some properties of objects in the universe outside our Solar System as well as theories about the origin of the entire universe.

Science Content Objectives
Students will collect evidence that:
1. Very bright light that is far away from the eye can appear the same as dimmer light that is closer to the eye.
2. Space contains many objects, including galaxies and nebulae.
3. Our Solar System is part of the Milky Way Galaxy.

Inquiry Process Skills
Students will:
1. Conduct investigations that model scientific processes.
2. Use the findings to interpret scientific phenomena.

Connections to Standards and Benchmarks
In **Investigation 7**, students extend what they have already learned by investigating objects in the universe beyond the Solar System. The observations and investigations they complete will help them understanding the National Science Education Standards and AAAS Benchmarks below:

NSES Links
- A system is an organized group of related objects or components that form a whole.
- Science assumes that the behavior of the universe is not capricious, that nature is the same everywhere, and that it is understandable and predictable. Students can develop an understanding of regularities in systems, and by extension, the universe; they then can develop understanding of basic laws, theories, and models that explain the world.
- Order — the behavior of units of matter, objects, organisms, or events in the universe — can be described statistically. Probability is the relative certainty (or uncertainty) that individuals can assign to selected events happening (or not

happening) in a specified space or time. In science, reduction of uncertainty occurs through such processes as the development of knowledge about factors influencing objects, organisms, systems, or events; better and more observations; and better explanatory models.

AAAS Links
- The universe contains many billions of galaxies, and each galaxy contains many billions of stars. To the naked eye, even the closest of these galaxies is no more than a dim, fuzzy spot.
- The Sun is many thousands of times closer to the Earth than any other star. Light from the Sun takes a few minutes to reach the Earth, but light from the next nearest star takes a few years to arrive. The trip to that star would take the fastest rocket thousands of years. Some distant galaxies are so far away that their light takes several billion years to reach the Earth. People on Earth, therefore, see them as they were that long ago in the past.

Teacher Commentary

Preparation and Materials Needed

Preparation
It may be useful to download and print out a set of photographs from the Hubble Space Telescope web site: http://hubblesite.org/gallery/. Choose photographs of galaxies and nebulae. If you print them out on heavy paper, you can laminate them for future use.

This investigation requires six or seven 40-minute class periods to complete, depending upon how you structure it. **Day One:** Have the students address the **Key Question** and review the kinds of ideas they have and record them in their Journals. They can then review the whole investigation and make plans on how to proceed. **Day Two:** Have students begin **Part A** of this investigation. **Day Three:** Have students complete **Part B** of the investigation. **Day Four:** Students can complete **Part C** of the investigation. **Day Five:** Students can complete **Part D** of the investigation. **Day Six:** Use this period for pulling the whole investigation together.

Suggested Materials
Part A
- small flashlight with a powerful beam
- large flashlight with a dim beam
- metric measuring tape

Part B
- star chart
- glow-in-the-dark stars
- heavy, black construction paper

Part C
- Hubble telescope photographs of galaxies and nebulae
- CD-ROM on the universe (optional)

Investigating Earth in Space: Astronomy

Investigation 7: What is Beyond Our Solar System?

Investigation 7:
What is Beyond Our Solar System?

Key Question
Before you begin, first think about this key question.

What are the other major objects in our universe and what are they like?

Think about what you have learned about the Solar System. What is beyond our Solar System?

Share your thinking with others in your group and in your class.

You have spent quite a bit of time in this module studying the objects in our own Solar System. You should now know about the role of the Sun, the names and characteristics of the planets (including the Earth), and the structure of the Solar System. You should also know the role of gravity in orbital paths, and how our Moon is related to and affects the Earth. In this investigation, you will move beyond our Solar System to learn about stars, galaxies, nebulae, and theories of how the universe was formed.

Investigate
Human beings have always been fascinated by stars and constellations. Constellations are the patterns stars make in the night sky. Knowing about the constellations and where they appear in the sky at different times of the year is useful in identifying individual stars.

Materials Needed
For this investigation, your group will need:
- star chart
- glow-in-the-dark stars
- large sheet of black construction paper
- small flashlight with a very powerful beam
- large flashlight with a dim beam
- metric measuring tape
- CD-ROM on the universe (optional)

Teacher Commentary

Key Question

By now, students should have a strong grasp of the Solar System. They may, however, be quite uncertain of the universe beyond. The **Key Question** is designed to help start them thinking about the universe.

Write the **Key Question** on the chalkboard or on an overhead transparency. Have the students record their answers in their journals. Ask them to think about and answer the question individually. Ask them to write as much as they know and to provide as much detail as possible in their responses. Emphasize that they should include the current date and the prompt (the question itself, a meaningful heading, etc.) in their journal entries.

Student Conceptions about the Universe

Unless they have a special interest, are close to people who have, or have studied astronomy in a planetarium, museum class, or for a scout badge, most students will only have a rudimentary understanding of objects beyond the Solar System. Their answers to the **Key Question** should reveal some of this. Commonly held ideas about the universe include:
- Stars and constellations appear in the same place in the sky every night.
- All the stars in a constellation are near each other.
- All the stars are the same distance from the Earth.
- The galaxy is very crowded.
- Stars are evenly distributed throughout the universe.
- All stars are the same size.
- The brightness of a star depends only on its distance from the Earth.
- Stars are evenly distributed throughout the galaxy.
- The constellations form patterns clearly resembling people, animals or objects.

Answer for the Teacher Only

Besides the Sun and the planets, our universe contains stars (glowing balls of burning gas); nebulae (clouds of gas and dust); asteroids (rocky and metallic bodies that orbit the Sun); meteoroids (rocky bodies that revolve around the Sun – smaller than asteroids); comets (small bodies of ice, rock or dust loosely packed together); and black holes (collapsed stars).

Investigate

Teaching Suggestions and Sample Answers

This investigation has four parts. Explain to students that they will be working in parallel groups and sharing their results and ideas after each section.

INVESTIGATING EARTH IN SPACE: ASTRONOMY

Part A: How Stars Look to Our Eyes

1. To get a sense of the relationship between the brightness of stars and their distance from Earth, your group will try a little demonstration. You will need a strong flashlight and a much weaker flashlight to do this.

2. Two members of your group will each take a flashlight. The other members will stand at a distance of at least 10 m away.

3. The students with the flashlights will turn on both flashlights, in a large darkened area. The observers will note the differences in the way that the flashlights appear.

4. As a group, brainstorm how the flashlights could be arranged so that they appear to be the same brightness to the observers. Try different ideas until the flashlights appear to be the same brightness to the observers.

   a) Record the steps you took and the end locations of each flashlight in your journals.

5. Answer the following questions in your journal:

   a) What did you have to do to make both lights look about the same?

   b) What might that tell you about the stars that you see in the night sky?

   c) Do you think all stars are the same distance from the Earth? What evidence do you have for or against this idea?

   d) If your class has access to computers with an Internet link-up, see what you can find out about stars and their distance from the Earth on the NASA web site: www.nasa.gov. Did this information match your group's ideas about stars and how they appear from the Earth?

   e) What else can you find from the NASA site about how stars appear to twinkle from the Earth? Find this out and share it with other groups in your class.

Inquiry
Initial Experiments

Often, a scientific investigation begins with a simple informal experiment to test a prediction. The results may not solve the problem, but they may be useful in later investigations. In this part of the investigation, you did a simple demonstration to start you thinking about whether all the stars you see at night are the same distance from the Earth.

Teacher Commentary

Part A: How Stars Look to Our Eyes

This demonstration will help your students to see that stars can be different sizes and that a star's relative brightness is not necessarily an indicator of how close or far away it is from an observer. This may be a new idea for some students. It is very easy to look up at the night sky and interpret the stars as all being the same distance from Earth. Visits to a planetarium could help to reinforce this inaccuracy for some students. It is also easy to think that the brighter the star, the closer it is. There's no obvious reference point against which to judge this. Be alert to your students' ideas about these topics and help them toward a better understanding of the concepts.

Before class, you will need to find the best place for your students to conduct the flashlight experiment. They will need an area they can darken somewhat, such as a hallway without windows or a theater.

5. a) The dimmer flashlight needs to move closer to the observer or the brighter one needs to be moved farther away to make them both look about the same.
   b) This shows that the way that the star appears from the Earth does not necessarily represent how far it is from the Earth.
   c) Student answers may vary. Students should recognize that not all starts are the same distance from the Earth. They appear brighter and dimmer to the naked eye.
   d) Student responses will vary.

> **Assessment Tools**
> **Investigation Journal Entry-Evaluation Sheet**
> Use this sheet to help students learn the basic expectations for journal entries that feature the write-up of investigations. It provides a variety of criteria that both you and your students can use to ensure that their work meets the highest possible standards and expectations. Adapt this sheet so that it is appropriate for your classroom, or modify the sheet to suit a particular investigation.

Investigating Earth in Space: Astronomy

Investigation 7: What is Beyond Our Solar System?

Part B: Your "Specialist" Constellation

1. Your teacher will give you a "map" of the night sky at a particular time of the year. Work with your group to decide on a constellation that you find really interesting. You may have heard about one in language arts studying mythologies of different cultures.

2. Once your group has agreed upon its constellation, find out the following information about it to share with other groups in your class.

   a) How did it get its name and is there any story behind that?

   b) What stars are in it?

   c) How does it appear to change its position with the seasons?

   d) What "famous" stars are in it?

3. Use glow-in the-dark stars to "make" your constellation on a sheet of black construction paper. You will also need to become aware of where this constellation is in relation to other constellations in the sky. Your teacher will later hang the constellation posters on the ceiling in their proper locations.

4. Once all the constellations are on the ceiling, turn off the classroom lights. Have a whole group session during which each group explains the facts about its constellation to the other groups.

   a) Make notes on the other constellations so that you can ask questions about them later on.

Inquiry
Using References

When you write a science report, the information you gather from books, magazines, and the Internet comes from evidence gathered by others. You must always list the source of your evidence. This not only gives credit to the person who wrote the work, but it allows others to examine it and decide for themselves whether or not it makes sense.

Teacher Commentary

Part B: Your "Specialist" Constellation

1. There are 88 officially recognized constellations, but they are not all visible from any particular location all of the time. In view of this, it may be helpful to limit the constellations groups will study to those that are more familiar. If possible, choose those that are visible in the night sky at the time students are working on this investigation. You can find out which are visible using the "Constellations by Month" guide on web sites, such as this:
http://www.astro.wisc.edu/~dolan/constellations/constellationmonth_list.html

2. The naming of constellations goes back at least 5000 years. Poets, farmers and early astronomers have all contributed to this over time. Your students should be able to find out a great deal about their chosen constellations: their links to ancient mythology through to today's popular notions, including astrology. What's important is that your students do not confuse fact with fiction. Be sure that they focus on the scientific aspects of this study in a scientific way.

3. Arrange the students' posters on the ceiling of the classroom after they have been completed. You may want to enlist the help of the students to orient the posters correctly with respect to other constellations.

4. Set aside enough time for all groups to talk about their constellations and answer questions.

INVESTIGATING EARTH IN SPACE: ASTRONOMY

Part C: Galaxies, Nebulae, and the Origin of the Universe

1. Examine the photographs of galaxies taken by one of the Great Observatories.

   a) Examine the pictures closely. What other galaxies do you see? How do the galaxies seem to be similar? How are they different?

   b) In your journal, make a table like the one below. As you study your galaxy photographs, fill in the table as best you can. You will also want to refer to the **Digging Deeper** on Galaxies and Nebulae to help you out.

Name of the Galaxy	Shape of the Galaxy	Important Stars in the Galaxy	Other Interesting Facts About the Galaxy

2. Now, examine the pictures that you have of nebulae (the plural of nebula) in the same way. You already know that nebulae are enormous clouds of dust and gas in the universe.

   a) How are the nebulae the same and how they are different?

   b) In your journal, make a table like the one below for nebulae. Complete the information, again, using the information on Galaxies and Nebulae for help.

Name of the Nebula	Shape of the Nebula	Type of Nebula	Other Interesting Facts About the Nebula

Teacher Commentary

Part C: Galaxies, Nebulae, and the Origin of the Universe

1. Have a general discussion about the properties and characteristics of galaxies with your students to avoid confusion between galaxies and the constellations from the last investigation. Remind students about the role gravity plays within galaxies.

2–3. In addition to drawing the tables in their journals, student groups could make large-scale versions for sharing and displaying later.

Investigation 7: What is Beyond Our Solar System?

3. When you finish, compare what you have discovered about galaxies and nebulae with another group.

   a) Share your information and make your tables as complete and accurate as you can. Keep these tables handy for the **Review and Reflect** questions at the end of this investigation.

Part D: Theories About How the Universe Began

Over the years, scientists have had many theories about how the universe began. It has always been very difficult, however, to collect evidence that would support these theories. With advancing technologies, though, scientists have been able to collect data that seem to support one or two major theories. In this last part of the investigation, you will search the **Digging Deeper** section for evidence supporting one of these, the "Big Bang" Theory.

Inquiry
Using References as Evidence

When you write a science report, the information you gather from books, magazines, and the Internet comes from science investigations. Just as in your experiments, the results can be used as evidence. Sometimes, enough new evidence accumulates that make ideas change drastically. This is true of the theories about how the universe began.

1. You will first need to find out what the Big Bang Theory is.

   a) What does it seem to explain about the universe and who came up with it in the first place?

2. Use the evidence about the Big Bang Theory in the **Digging Deeper** section.

   a) Work with your group to write an argument supporting the Big Bang Theory.

3. When you finish, share your argument with the class. Listen to what other groups say about the theory. They may have spotted evidence that your group missed.

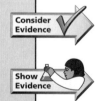

   a) How complete, do you think, is the evidence of the Big Bang Theory?

   b) What appears to be missing in the evidence?

   c) How much sense does the evidence make to you? How useful is the theory in thinking about the universe?

Investigating Earth Systems

A 59

Part D: Theories About How the Universe Began

2. Students should be encouraged to use resources in addition to the **Digging Deeper** section as a basis for constructing their supporting argument for the "Big Bang" theory. Since this popular theory is still under critical scrutiny, up-to-date web-based information from scientifically reliable sources may be especially useful. This should ensure a lively discussion later.
3. Make sure students have a clear understanding of these questions before they start gathering their evidence in support of the Big Bang theory. Let them know that their group will be answering these questions in the class discussion. Stress the need for finding and sharing scientific evidence in this investigation.

Investigating Earth in Space: Astronomy

INVESTIGATING EARTH IN SPACE: ASTRONOMY

Digging Deeper

GALAXIES AND NEBULAE

As You Read...
Think about:
1. What are galaxies and how are they classified?
2. Where are we in the Milky Way Galaxy?
3. What do scientists think "black holes" are?
4. What is the "Big Bang Theory"? What is the evidence for this theory?

Galaxies

Galaxies are large systems of stars, nebulae, and the matter between the stars (interstellar matter). A number of galaxies were discovered and cataloged by Charles Messier in the late 1700s. Messier's telescope did not have the resolution necessary to see individual stars in the galaxies and he referred to them as nebula.

Milky Way Galaxy (spiral): This is our own galaxy. It is 100,000 light-years in diameter.

Andromeda Galaxy (spiral): This is a relatively close spiral galaxy similar to our own (2–3 light-years away).

M84 (lenticular): Sixty million light-years away.

Virgo A or M87 (giant elliptical): Sixty million light-years away.

Types of Galaxies

- **Spiral:** These galaxies have a large central disc with clusters of young stars and lots of matter between the stars and a bulge component of older stars.

- **Lenticular:** These galaxies are "smooth disc" galaxies of older stars. They have used up the material between the stars.

- **Elliptical:** These galaxies are football-shaped galaxies of older stars with little or no material between the stars.

Teacher Commentary

Digging Deeper
This section provides text, photographs, and illustrations that give students greater insight into the topic of the universe. You may wish to assign the **As You Read** questions as homework to help students focus on the major ideas in the text.

As You Read...
Think about:
1. Galaxies are large systems of stars, nebulae, and the matter between the stars. They are classified by their shape and can be spiral, elliptical, lenticular, or irregular.

2. Our Sun and Solar System are located on the outside of one of the arms of the spiral galaxy.

3. For a long time, scientists thought that black holes formed when stars collided as matter moved towards the center of the galaxy during its beginnings. However, more recent findings (2000) support that "At least 15 percent of supermassive black holes have formed since the Universe was half its present age." (Barger, A., 2002). The black holes produce "active" galaxies where there is more energy being emitted than would normally be expected.

4. The Big Bang Theory explains the origin of the universe. The evidence is that the material within the universe is still expanding.

Galaxies and Nebulae
This section will help your students to understand similarities and differences between galaxies in the universe. They will be able to compare distant galaxies with our own Milky Way.

The Milky Way Galaxy
From reading this section, students will be able to get a better understanding of the galaxy within which our Solar System exists. It's important that students begin to appreciate the enormity of the universe and also how scientists have come to know what they know about it.

> **Assessment Opportunity**
> You may wish to rephrase selected questions from the **As You Read** section into multiple choice or true/false format to use as a quiz. Use this quiz to assess student understanding and as a motivational tool to ensure that students complete the reading assignment and comprehend the main ideas.

Investigation 7: What is Beyond Our Solar System?

- **Irregular:** These galaxies are those that don't fit into the other categories.

Nebula

A nebula is an enormous cloud of gas and/or dust in space. Nebulae are the birthplaces of stars.

Types of Nebulae

- **Emission Nebulae:** These are clouds of gas, which glow by re-emitting the ultra-violet radiation absorbed from young hot stars. These nebulae usually look reddish and are the source of recent star formation.
- **Reflection Nebulae:** These are clouds of dust that reflect the light of nearby stars. They usually look blue and are also the source of star formation.
- **Dark Nebulae:** These are usually about a few hundred light-years in width. Dark nebulae are clouds of dust that block the light from behind them.
- **Planetary Nebulae:** These are relatively small clouds of dust given off by dwarf stars as they near the end of their lives.
- **Supernova Remnants:** These are a relatively small (few light-years across) part of a massive star that is left over after the star ends its life in a supernova explosion.

Stars are formed from giant nebula clouds.

Horsehead Nebula (dark nebula at the center).

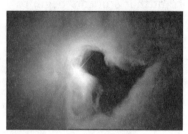
NGC 1999, a nebula in the constellation Oriion. (reflection)

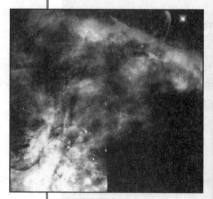
Orion Nebula (emission).

Teacher Commentary

About the Photo

The photo on the bottom right of page A61 of the student text shows the Orion Nebula. The Orion Nebula is the site closest to the Earth which shows massive star formation. The nebula is an enormous cloud of gas surrounding a cluster of very hot young stars and is located within the constellation Orion. The photo on the bottom left on page A61 of the student text shows the star NGC 1999. Scientists believe this is a very young star, that it may be only about 5000 years old.

INVESTIGATING EARTH IN SPACE: ASTRONOMY

The Milky Way.

The Milky Way Galaxy

Our Solar System is located in a galaxy known as the Milky Way Galaxy. The Milky Way Galaxy contains more than 100 billion stars, each of which may have orbiting planets and other objects. The Milky Way Galaxy is shaped like a spiral, with arms that extend outwards from a bulge at its center. Each arm is full of dust, stars, and space. Our Solar System is located on the outside of one of these arms.

The arms of the Milky Way Galaxy rotate on an axis that goes through the middle of the galaxy. This means that just as the planets in our Solar System orbit the Sun, our Solar System orbits around the center of the Milky Way. It takes 240 million years for the Sun to orbit the center of the Milky Way Galaxy!

Scientists believe that many galaxies contain gigantic black holes in their centers. These black holes formed when stars collided as matter moved toward the center of the galaxy during its beginnings. The black holes produce "active" galaxies where there is more energy being emitted than would normally be expected.

Teacher Commentary

About the Photo

The photo on page A62 of the student texts shows the Milky Way Galaxy. Our Sun and the planets that orbit it are all part of the Milky Way, a spiral galaxy. The Milky Way is estimated to contain at least 200 billion stars and has a diameter of about 100,000 light-years.

Investigation 7: What is Beyond Our Solar System?

The gravitational pull of a black hole is so intense that nothing can escape, not even light.

In contrast, "normal" galaxies give off energy from their stars without this additional source. Change takes place much more slowly in normal galaxies than in active ones. The Milky Way Galaxy is a normal galaxy.

THE UNIVERSE BEGAN WITH A BANG! (OR DID IT?)

The Big Bang Theory is the most widely accepted explanation among scientists for the origin of the universe. It states that the universe was formed 13.7 billion years ago. In the beginning, everything in the universe was concentrated into a volume that was incredibly small. The matter that made up this tiny universe was very hot and very dense (heavy for its size). The early universe began with space rapidly

Our Solar System is located in a spiral arm of the Milky Way Galaxy. The stars in the inner bulge of the spiral were the first to form when the Milky Way Galaxy began to develop. Stars in the arms of the spiral formed later and are younger than those in the bulge.

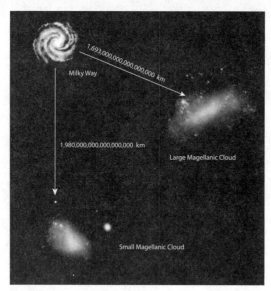

The distances between galaxies are great.

Teacher Commentary

The Universe Began With A Bang! (Or Did It?)

This is a complex and difficult idea for students (and scientists!) to come to grips with. How things began is both an intriguing and mysterious question. Keep in mind that some students (and their parents) may have strong religious views about the origins of the universe, which sharply differ from the commonly accepted scientific explanations.

Assessment Opportunity

You may wish to rephrase selected questions from the **As You Read** section into multiple choice or true/false format to use as a quiz. Use this quiz to assess student understanding and as a motivational tool to ensure that students complete the reading assignment and comprehend the main ideas.

About the Photo

The photo at the top of page A63 of the student text shows a black hole. A black hole is a region of space that has so much mass concentrated in it that there is no way for a nearby object to escape its gravitational pull. Black holes can't be seen directly, therefore scientists rely on indirect evidence that black holes exist.

INVESTIGATING EARTH IN SPACE: ASTRONOMY

expanding carrying matter along with it. The Big Bang Theory also states that the universe is still expanding. Distant galaxies are traveling away from each other at great speeds.

Evidence of the Big Bang

The Big Bang Theory explains only the expansion of the universe. It does not attempt to explain how this process began. Nor does it explain what is beyond the edge of the universe. As with many strong theories, its strength comes from making a clear statement that is supported by solid evidence.

Stronger telescopes in the 20th century helped many astronomers investigate faraway galaxies and the Big Bang Theory. In 1929, Edward Hubble discovered that the more distant a galaxy is from our galaxy, the faster it is moving away from us. Specialized astronomers who study the origin and expansion of the universe are known as cosmologists.

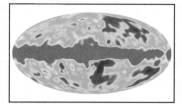
A microwave map of the Universe produced by the Wilkinson probe.

Cosmologists study the universe by using probes that look out into space. These probes detect the energy released during the initial moments of the Big Bang. This energy has been detected by these probes coming in from all directions of the universe.

By studying the light from distant galaxies and the energy from the edges of the universe, scientists are actually looking back through time.

The Wilkinson probe mapping the edge of the universe.

Teacher Commentary

About the Photo

The photo on page A64 of the student text shows the Wilkinson Microwave Anisotropy Probe. This probe was sent to produce high-fidelity, all-sky, polarization-sensitive maps that can be used to study the cosmic microwave background. It is believed that such a map will provide answers to questions about the origin and ultimate fate of our universe.

Investigation 7: What is Beyond Our Solar System?

A STAR IS BORN

Astronomers believe that a star begins to form when particles in a dense region of a gas and dust cloud nebula are pulled toward each other. This collapse is caused by gravitation as the particles move inwards toward the center of the nebula to form the star. Eventually, enough particles collect to make the star dense enough to produce energy. Most of the energy produced is by hydrogen nuclei joining together, deep in the star's center to form helium nuclei. This process is called hydrogen fusion.

Main-Sequence Stars

The diagram shows the relationship between a star's brightness and its temperature. The stars in the diagram are categorized into one of three types: white dwarfs, main-sequence stars, and supergiants and giants. The stars in the main-sequence section show that as temperature increases, brightness increases. About 90% of the stars in the universe fit into this sequence. The

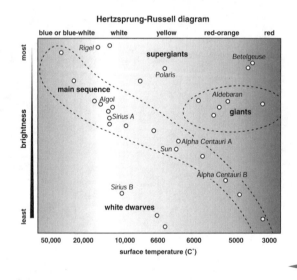

Teacher Commentary

A Star is Born

Have students use this as a reference point when they are considering their own adopted constellation or galaxy. By now they should fully understand that stars are suns, similar to our own. This section, which covers **The Color of Stars, Main-Sequence Stars,** and **The Life Cycle of a Star,** will be useful background for their investigations.

Encourage students to search for other sources of information in addition to this **Digging Deeper** section. They can consult reference books, video programs, such as those offered by the Discovery/Science Channel and National Geographic Society. There is a wealth of information available on the Internet, both governmental (NASA) and educational.

INVESTIGATING EARTH IN SPACE: ASTRONOMY

diagram also shows that as stars become cooler, they also become dimmer. As their brightness *decreases*, they change color from bright blue to dim red. The white dwarfs are small stars that are very hot, but not very bright. Supergiants and red giants are brighter than the hottest main-sequence stars. These stars are not as hot as the blue stars in the sequence.

The Life Cycle of a Star

The temperature of stars is so high that they cause nuclei of hydrogen to fuse together to create helium atoms in a process called fusion. For this reaction to occur, four hydrogen nuclei must combine to form a single helium nucleus. During this reaction, a small amount of mass is lost and converted into a massive amount of energy. All stars have a limited supply of hydrogen that can be fused into helium. Once the hydrogen is used up, the star goes through some big changes.

A picture of a giant galactic nebula, showing various stages of the life cycle of stars in one single view.

When main-sequence stars use up their hydrogen atoms, they become giant stars. It can take millions to billions of years for a star to deplete its supply of hydrogen. When a main-sequence star uses up its hydrogen supply, it starts to cool. As it cools, it begins to contract and become smaller. The contraction eventually makes the star heat up again causing the star to expand. During this process, its outer layers become much cooler than when it was a main-sequence star. The star is now like a car that is running out of gas. In its next cycle, it swells in size to become a giant star. Our Sun will

Teacher Commentary

NOTES

Investigation 7: What is Beyond Our Solar System?

Impressions of supernova explosions in neighboring galaxies.

use up its supply of hydrogen and go through this process in about 5 billion years.

White dwarfs form after the hydrogen in a star's core is used up. The star starts to cool and contract and it loses its outer layers into space. Gravity continues to draw matter toward the core of the star and it becomes a hot, very dense star of low brightness.

Giant stars are more than ten times larger than the Sun and can go through violent explosions. In the super-hot cores of giant stars, some of the matter fuses together. The star expands rapidly to a gigantic size. Supergiants form from giant stars and are much larger than the Sun. Eventually, iron forms in the core and energy production comes to an end. Once their hydrogen supply is depleted, their high core temperatures cause violent reactions to occur. The core collapses violently, sending shock waves through the star. This creates a gigantic explosion called a supernova.

Teacher Commentary

About the Photo

The photo on page A67 of the student text shows a supernova explosion. A supernova explosion will occur when there is no longer enough fuel for fusion in the core of the star to create an outward pressure which combats the inward gravitational pull of the star's great mass. First, the star will swell into a red supergiant. The core of that star yields to gravity and begins shrinking. As it shrinks, it grows hotter and denser. When the core contains essentially just iron, it has nothing left to fuse. Fusion in the core ceases. In less than a second, the star begins the final phase of gravitational collapse. The core temperature rises to over 100 billion degrees. The core compresses, but then recoils. The energy of the recoil is transferred to the envelope of the star, which then explodes and produces a shock wave. As the shock encounters material in the star's outer layers, the material is heated, fusing to form new elements and radioactive isotopes. The shock then propels the matter out into space. The material that is exploded away from the star is now known as a supernova remnant.

INVESTIGATING EARTH IN SPACE: ASTRONOMY

HOW DO SCIENTISTS INVESTIGATE OUTER SPACE?

The invention of the telescope during the time of Galileo was one of the most important events in the history of astronomy. This small instrument completely changed how people thought about the stars, planets, and moons in space. It was an extension of human senses and, for the first time, people could see things never before dreamed of. Since Galileo's time, countless telescopes have been designed, each one providing clearer images of objects in space.

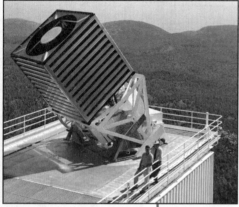

The 2.5-meter reflecting telescope of the Sloan Digital Sky Survey. The box-like structure protects the separately mounted telescope from being buffeted by the wind.

Today's optical telescope magnifies objects in space, such as stars and planets, by concentrating the visible light waves they emit or reflect. These telescopes use lenses or mirrors to gather the light from an object. The light is then focused to create a magnified image of that object. There are optical telescopes that are small enough to be carried in your hand, while others are as big as buildings and weigh 300 tons!

The Zeiss 12-inch refracting telescope

Teacher Commentary

About the Photo

The photo at the top of page A68 of the student text shows a 2.5- meter reflecting telescope at the Sloan Digital Sky Survey. The Sloan Digital Sky Survey observing site is located at Apache Point Observatory in the Sacramento Mountains of New Mexico. The site of the telescopes is located about 9150 feet above sea level. The combination of height and distance from civilization makes for dark nights with clean air and great visibility. The Sloan Digital Sky Survey will undertake one of the most massive projects of all time. It is designed to systematically map one-quarter of the entire sky, producing a detailed image of it and determining the positions and absolute brightness of more than 100 million celestial objects. Scientists with the survey will also measure the distance to a million of the nearest galaxies. This will provide a three-dimensional picture of the universe. The Sky Survey will also record the distances to 100,000 quasars.

The photo at the bottom of page A68 of the student text shows the Zeiss 12-inch refracting telescope at the Griffith Observatory in Los Angeles, CA. This telescope is used to show the general public such distant objects as the Moon, planets, and the brightest showpiece objects of our galaxy.

Investigating Earth in Space: Astronomy

Investigation 7: What is Beyond Our Solar System?

The Earth's atmosphere can get in the way of the visible light gathered by optical telescopes. The gases, clouds, and particles in the atmosphere can make the visible light from stars appear blurry or flicker. To prevent this from happening, scientists put their telescopes on mountaintops, where the air is dryer and thinner. There is also less pollution to distort the visible light coming from space.

The Hubble Space Telescope (HST) orbiting above Earth.

In 1990, scientists put the Hubble Space Telescope into orbit above the Earth's atmosphere. In space, there is little to distort the visible light coming from distant objects. It has produced some of the most detailed and clearest images of space ever seen. Hubble is about the size of a large school bus. It travels at an orbital speed of nearly 8 km/s (kilometers per second), or 97 min per orbit. Hubble is powered by energy from the Sun and in an average orbit, uses about the same amount of energy as twenty-eight 100-W light bulbs.

Teacher Commentary

About the Photo

The photo on page A69 of the student text shows the Hubble Telescope in orbit around the Earth. The Hubble Telescope was launched into orbit in 1990 and has continued to explore new worlds beyond Earth. Recently, the Hubble Telescope has provided data and images of dark energy, gamma-ray bursts, and information on the atmospheres of distant planets.

INVESTIGATING EARTH IN SPACE: ASTRONOMY

The Arecibo Observatory located in Puerto Rico.

Other telescopes can collect different types of electromagnetic radiation from objects in space. The largest of these are the radio telescopes. They have huge surfaces to receive long radio wavelengths from space objects. Arecibo is the largest radio telescope in the world with its 300-m wide surface built into a hill! There are also ultraviolet telescopes, infrared telescopes,

The W. M. Keck Observatory on the summit of Hawaii's dormant Mauna Kea volcano. The twin Keck Telescopes are the world's largest optical and infrared telescopes.

Teacher Commentary

About the Photo

The photo at the top of page A70 of the student text shows the Arecibo Telescope located in Puerto Rico. The Arecibo Observatory is part of the National Astronomy and Ionosphere Center, a national research center operated by Cornell University. The Arecibo Telescope is the world's largest single-dish radio telescope.

The photo at the bottom of page A70 of the student text shows the Keck telescope in Mauna Kea, Hawaii. The Keck telescope is operated by the California Association for Research in Astronomy.

Investigation 7: What is Beyond Our Solar System?

gamma-ray telescopes, and x-ray telescopes. They work in space and collect radiation from above the Earth's atmosphere. Each of these telescopes creates its own image for the data that it is designed to collect.

A human-made object placed in orbit around the Earth is called an artificial satellite. Artificial satellites serve a variety of purposes, including the transmission of signals for television shows and cell phone calls. When a telescope is placed in space above the Earth's atmosphere, it is also an artificial satellite.

Scientists also send research instruments, such as telescopes, that travel into space away from the Earth. These unmanned spacecraft are known as space probes. The first probe was launched in 1959 to collect information about the Earth's Moon. Since that time, dozens of probes have been sent into space to collect information about all the planets of the Solar System, as well as asteroids, comets, and solar wind. Probes have even landed on the surface of Mars!

The first single crewmember EVA capture attempt of the Intelsat VI as seen from Endeavour's aft flight deck windows. EVA Mission Specialist Pierre Thuot standing on the Remote Manipulator System (RMS) end effector platform, with the satellite capture bar attempting to attach it to the free floating communications satellite.

The Mars Pathfinder lands on Mars. The Sojourner used the fully deployed forward ramp at far left, and rear ramp at right, to descend to the surface of Mars on July 5, 1997. Rover tracks lead to Sojourner, shown using its Alpha Proton X-Ray Spectrometer instrument to study the large rock Yogi.

Teacher Commentary

About the Photo

The photo at the top of page A71 of the student text shows a Landsat Satellite. The Landsat program is designed to gather information and images of the Earth from space. The first Landsat satellite was launched into space in 1972 and the most recent, Landsat 7, was launched in 1999. The images gathered are a unique resource for global change research and applications in agriculture, geology, forestry, regional planning, education and national security.

The photo on the bottom of page A71 of the student text shows an image of the surface of the planet Mars. The Mars Exploration Rovers landed on the surface of Mars in January 2004. This mission is part of NASA's Mars Exploration Program, a long-term effort of robotic exploration of the red planet. The goals of the mission are to search for clues to potential water activity on Mars. If evidence of water were definitively found, then this could open the door to the possibility that life once existed on the Red Planet.

INVESTIGATING EARTH IN SPACE: ASTRONOMY

Review and Reflect

Review

1. Explain why you think that stars, although they are incredibly large, can look so tiny to us here on the Earth. Why does our own star, the Sun, look so big?

2. What are constellations? How can you explain how they look different in the sky at various times of the year?

3. What are galaxies? How do scientists classify them?

4. Explain what nebulae are and how they are different from one another.

5. What is the Big Bang Theory? What is one piece of evidence supporting that theory?

Reflect

6. What do you think are the main problems in studying objects in the universe?

7. How has technology helped scientists to learn more about what is in the universe? What new technology do you think would be useful in studying the universe?

8. How well do you think the Big Bang Theory is supported? How confident are you that this theory explains how the universe began?

Thinking about the Earth System

9. What conditions would have to be in place for a planet similar to Earth to exist?

Thinking about Scientific Inquiry

10. In which parts of the investigation did you:
    a) Make models?
    b) Compare ideas?
    c) Revise your ideas?
    d) Use your imagination?
    e) Share ideas with others?
    f) Organize information?
    g) Pull your information together to make a presentation?

Evidence for Ideas

Teacher Commentary

Review and Reflect

Review
These five questions will help students to review their work systematically. Set aside enough time for them to do this carefully. They have been dealing with some difficult concepts and will probably need help to understand the theories.

1. The distance a star is from Earth determines how we will see it. The Sun looks large because it is the closest star to our planet. The other stars are so far away that they look very small to us.
2. Constellations are patterns that stars make in the sky. The Earth is rotating on its axis and moving through space, as are the stars. This is what makes the constellations appear to change locations.
3. Galaxies are large systems of stars, nebulae, and the matter between the stars. They are classified by their shape and can be spiral, elliptical, lenticular, or irregular.
4. A nebula is an enormous cloud of gas and/or dust in space. Nebulae are the birthplaces of stars.
5. The Big Bang Theory is the most widely accepted explanation among scientists for the origin of the universe. It states that the universe was formed sometime between 12 and 14 billion years ago.

Reflect
In the same way, the reflective questions should guide students toward a better understanding of these broad issues.

6. Answers will vary but students should recognize that it is the great distances that often make objects difficult to study.
7. Technology has allowed scientists to look farther into the universe and collect more data from distant objects. Computers, telescopes, and improved space probes can be useful in gathering information.
8. Answers will vary. Be sensitive to personal and religious beliefs.

Thinking about the Earth System
9. Very similar conditions to Earth would have to be in place, in particular the interactions of geosphere, atmosphere, hydrosphere and biosphere. The planet would have to have a similar relationship with a sun's gravitational pull and heat energy and have similar tectonic activity. For life to exist, the planet would need a similar balance of oxygen, carbon dioxide, water, and temperature.

Thinking about Scientific Inquiry
Make sure students track which processes they have and have not used in this investigation. Make sure that the full list of **Inquiry Processes** (**Blackline Master Astronomy 1.2**) is available while they are doing this.

> **Assessment Opportunity**
> **Review and Reflect Journal-Entry Evaluation Sheet**
> Use the general criteria on this evaluation sheet for assessing content and thoroughness of student work. Adapt and modify the sheet to meet your needs. Consider involving students in selecting and modifying the criteria for evaluating their reflections on **Investigation 7**.

Teacher Review

Use this section to reflect on and review the investigation. Keep in mind that your notes here are likely to be especially helpful when you teach this investigation again. Questions listed here are examples only.

Student Achievement

What evidence do you have that all students have met the science content objectives?

_____

Are there any students who need more help in reaching these objectives? If so, how can you provide this? _____

_____

What evidence do you have that all students have demonstrated their understanding of the inquiry processes? _____

_____

Which of these inquiry objectives do your students need to improve upon in future investigations? _____

_____

What evidence do the journal entries contain about what your students learned from this investigation? _____

_____

Planning

How well did this investigation fit into your class time? _____

_____

What changes can you make to improve your planning next time? _____

_____

Guiding and Facilitating Learning

How well did you focus and support inquiry while interacting with students?

_____

What changes can you make to improve classroom management for the next investigation or the next time you teach this investigation? _____

_____

Teacher Review

How successful were you in encouraging all students to participate fully in science learning? _____

_____

How did you encourage and model the skills values, and attitudes of scientific inquiry? _____

_____

How did you nurture collaboration among students? _____

_____

Materials and Resources

What challenges did you encounter obtaining or using materials and/or resources needed for the activity? _____

_____

What changes can you make to better obtain and better manage materials and resources next time? _____

_____

Student Evaluation

Describe how you evaluated student progress. What worked well? What needs to be improved? _____

_____

How will you adapt your evaluation methods for next time? _____

_____

Describe how you guided students in self-assessment. _____

_____

Self Evaluation

How would you rate your teaching of this investigation? _____

_____

What advice would you give to a colleague who is planning to teach this investigation? _____

_____

NOTES

Teacher Commentary

INVESTIGATION 8: DISCOVERING THE DIFFERENCE BETWEEN SCIENCE FACT AND SCIENCE FICTION

Background Information

Science fiction (SF) can be described as a sub-category of a type of fiction broadly known as literature of the fantastic. Others include horror and fantasy. H. Bruce Franklin, writer and professor at Rutgers University, distinguishes it from other fiction in the following way:

> "On one side lies *fantasy*, the realm of the *impossible*. On the other side lie all the forms of fiction that purport to represent the actual, whether past or present. Science fiction's domain is the *possible*. Its territory ranges from the present Earth we know out to the limits of the possible universes that the human imagination can project, whether in the past, present, future, or alternative time-space continuums. Therefore, science fiction is the only literature capable of exploring the macrohistory of our species, and of placing our history, and even our daily lives, in a cosmic context."

Science fiction has antecedents all the way back to ancient times. For example, Homer mentions mechanical servants akin to robots in *The Iliad* and Lucian of Samosata (born c. A.D. 125) wrote a number of satirical dialogues based on fantastic ideas, making him perhaps the first writer of interplanetary fiction. However, as a body of literature, along with movies, art, comics, radio and TV shows, TV video and computer games, virtual reality, and others, SF is a relatively new phenomenon.

The word *science* is important. As the name suggests, science fiction is rooted in the so-called *scientific method*. It is also new. The word scientist has only been with us a short time appearing for the first time in 1840, and the term *science fiction* was first introduced a little over 150 years ago in 1851. As we know it today, science fiction literature falls into a range of aspects of the *fantastic* such as voyages of discovery, mythical adventure, and utopian societies. What makes these stories SF is the inclusion of science and technology into the fantastic mix. While SF authors often extend and explore futuristic science and technology possibilities, sometimes in a highly complex manner, they usually abide by the convention that any known scientific facts or processes should not be distorted or misrepresented.

The nature of science fiction lends itself neatly to a comparison to actual science. Indeed, some past science fiction ideas have become reality, or close to it. Jules Verne's *From the Earth to the Moon*, written in 1865, foreshadowed NASA's Apollo space program, and his submarine Nautilus in *Twenty Thousand Leagues under the Sea*, published in 1869, predated the first successful powered submarine by a quarter of a century. The first all-electric submarine, built in 1886 by two Englishmen, was named Nautilus in honor of Verne's vessel. The first nuclear-powered submarine, launched in 1955, was also named Nautilus. Blending science fact and fiction is the hallmark of science fiction. The ability to distinguish each from the other requires, at the very least, an understanding of science.

More Information…on the Web
Visit the *Investigating Earth Systems* web site www.agiweb.org/ies for links to a variety of other web sites that will help you deepen your understanding of content and prepare you to teach this module.

Teacher Commentary

Investigation Overview
In this, the final investigation, students are asked to distinguish between science fact and science fiction. As they complete this investigation, students will be challenged to show a clear understanding about what is science fact and what is science fiction and how the two can intertwine in a creative yet believable way. They are asked to produce a science fiction communication piece in which they must weave fiction and fact together. Creativity is stressed in this investigation. In this way their product, which forms the basis of student assessment, should demonstrate their level of understanding of astronomy concepts.

Goals and Objectives
The purpose of this activity is to review the content and inquiry processes that have been used throughout the module. It can be used as a final assessment, review for a final test, or both. As a result of completing **Investigation 8**, students will develop a better understanding of astronomy. They will also improve their ability to communicate science ideas and information to others.

Science Content Objectives
Students will demonstrate that they have:
1. A clear understanding of astronomical concepts and processes.
2. An ability to distinguish between science fact and science fiction based on evidence.
3. An ability to weave science fact and science fiction together to produce a convincing communications piece.

Inquiry Process Skills
Students will:
1. Conduct investigations that model scientific processes.
2. Collect and review both scientific information and science fiction techniques.
3. Work as a team to create and design a captivating communication piece.
4. Distinguish between commonly accepted scientific concepts and practices and pseudo-scientific fiction ideas.

Connections to Standards and Benchmarks

All the content standards and benchmarks students have been working to understand come together in this final investigation. Remember, these are statements of what students are expected to understand by the time they complete the eighth grade. What they have been doing throughout this module on Earth in space is just part of that ultimate learning outcome. Your students will have developed their understanding of some of these ideas, at least in part, but some students may require additional experiences.

As your students work through **Investigation 8**, keep the standards and benchmarks in mind and note the general level of understanding evident in what students discuss and do. Be especially alert to any confusion that a simple question from you might clarify, but do not attempt to teach these standards directly. Your role here is to guide students from the ideas they have toward a more complete understanding.

Teacher Commentary

Preparation and Materials Needed

Preparation

This investigation may vary according to the level of sophistication students can reach and the amount of time available for this investigation. In general, a minimum of five 40-minute class periods is desirable to complete the investigation, depending upon how you structure it. **Day One:** Have the students address the **Key Question** and review initial ideas they have and record them in their Journals. They can then review the entire investigation and make first plans on how to proceed. **Day Two:** Have students begin the investigation. **Day Three, Day Four:** Have students work on and develop their communication piece. **Day Five:** Have students present their communication pieces to each other, **Review and Reflect** on this final investigation, and review the whole module.

You and your students will have gathered much information and data that are relevant for this investigation from earlier investigations. You will also need to collect samples of the science fiction genre (movies, pictures, books, comic strips, etc.). It may be useful to have some of the classic science fiction novels available, such as: Jules Verne's *Journey to the Center of the Earth*, or *From the Earth to the Moon*; H.G. Wells' *Time Machine* or *War of the Worlds*; and *A Space Odyssey* by Arthur C. Clarke. Encourage students to bring in any science fiction materials they may have at home. They can look through these collected items to give them a sense of the genre and as a basis for creative ideas. You might find it useful to collaborate with your students' language arts teachers for this last investigation.

Be clear with all students about how they will be assessed in this final investigation. It is critical that they fully understand what is expected of them and exactly how their work will be evaluated.

This investigation may vary according to the level of sophistication students can reach and the amount of time available for this investigation. In general, a minimum of five 40-minute class periods is desirable to complete the investigation, depending upon how you structure it. **Day One:** Have the students address the **Key Question** and review initial ideas they have and record them in their Journals. They can then review the entire investigation and make first plans on how to proceed. **Day Two:** Have students begin the investigation. **Day Three, Day Four:** Have students work on and develop their communication piece. **Day Five:** Have students present their communication pieces to each other, **Review and Reflect** on this final investigation, and review the whole module.

Suggested Materials
- communications piece that combines astronomy fact with fiction (teacher's choice)
- materials to develop your own communications piece (video camera, PowerPoint™, computer, poster board, markers)

Investigating Earth in Space: Astronomy

Investigation 8: Discovering the Difference Between Science Fact and Science Fiction

Investigation 8:
Discovering the Difference Between Science Fact and Science Fiction

Key Question
Before you begin, first think about this key question.

How can you tell the difference between science fact and science fiction?

Think about what you have learned so far about Earth and space. Think about the science fiction movies or television shows you have watched. How are they different?

Share your thinking with others in your class.

There are many ways of communicating information about science. There are scientific journals, conferences, books and web sites. However, these are not the only ways people learn about science. In this investigation, you will first explore ways in which people communicate scientific information. Then you will practice separating science fact from science fiction.

Materials Needed
For this investigation, your group will need:

- communications piece that combines astronomy fact with fiction (teacher's choice)
- materials to develop your own communications piece (video camera, PowerPoint™, computer, poster board, colored pencils)

Teacher Commentary

Key Question

By now, your students should have a clear understanding of many important astronomical concepts. They may be uncertain about some, but will probably know how to find the information they need.

Write the **Key Question** on the chalkboard or on an overhead transparency. Have the students record their answers in their journals. Allow them to explore the scope of this question and have individual students contribute to the discussion from their experiences of the science fiction genre.

Student Conceptions

Most students will have some ideas and experience of science fiction. Distinguishing fact from fiction may be more difficult for them, especially when interpreting compelling images from movies and television programs. The science content and processes they have learned from this module will provide a basis for sifting scientific fact from fiction.

Answer for the Teacher Only

Scientific fact can be verified by accurate data, whereas science fiction cannot. Refereed scientific journals are important tools in differentiating between science fact and fiction, as are interviews with recognized experts in various scientific disciplines. It is not unusual for science fiction to anticipate science fact. For example, Jules Verne wrote about space travel and submarine travel long before they became realities. It's helpful to be aware of how science fiction is defined. Here are some examples:

- *Science Fiction is that class of prose narrative treating of a situation that could not arise in the world we know, but which is hypothesized on the basis of some innovation in science or technology, or pseudo-technology, whether human or extraterrestrial in origin.*
  —Kingsley Amis

- *Science fiction is really sociological studies of the future, things that the writer believes are going to happen by putting two and two together.* —Ray Bradbury

- *The best definition of science fiction is that it consists of stories in which one or more definitely scientific notion or theory or actual discovery is extrapolated, played with, embroidered on, in a non-logical, or fictional sense, and thus carried beyond the realm of the immediately possible in an effort to see how much fun the author and reader can have exploring the imaginary outer reaches of a given idea's potentialities.* —Groff Conklin

- *A handy short definition of almost all science fiction might read: realistic speculation about possible future events, based solidly on adequate knowledge of the real world, past and present, and on a thorough understanding of the nature and significance of the scientific method.* —Robert A. Heinlein

For more go to: http://www.panix.com/~gokce/sf_defn.html

Keep in mind that science fiction is different from fantasy in that it always has some relation to scientific fact or principles.

INVESTIGATING EARTH IN SPACE: ASTRONOMY

Investigate

1. With your group, think about and make a list of all the ways you find out about science information every day. Think creatively! There are a lot of choices!

   a) Write your list in your journal.

2. When you finish with your list, share it with other groups in your class to make a master list.

   a) On the master list, work with your class to make a check mark by any of the items on the list that might have science fact mixed with science fiction.

   b) Be sure to explain why you think that item on the list should be checked. It would help to give a specific example to make your explanation stronger.

3. Your group's task is to create an interesting and creative presentation for the general public on an astronomy topic you have learned about in this module.

   You may present it in any form, such as a comic book, TV show, commercial, movie trailer, but be sure to mix fact and fiction in your piece.

   When you finish, you will present your piece. The job of the other groups in the class will be to figure out which astronomy information is fact and which is fiction.

   You might want to focus on some of the basic science you learned about astronomy, or you might prefer to deal with some of the exciting new findings coming out of the field.

4. To practice how to do this, you will first analyze astronomy fact versus fiction in a movie, story, television show, or some other medium.

   a) As you watch or read the communications piece, make notes on what you think is scientific fact and what appears to be science fiction.

5. When you finish, discuss your notes with your group. Sort the fact from fiction, and then meet with another group to talk over your thoughts. Answer these questions:

Which photograph represents science fact and which is science fiction?

Teacher Commentary

Investigate

Teaching Suggestions and Sample Answers

This investigation is very largely student led. You need to be more adviser than "teacher." Do your best to help students realize their ambitions with this investigation. They are being asked to be creative, so try to respond to their ideas with encouragement and enthusiasm. Naturally, their ideas may go beyond what is possible, for resources, time, or other reasons. If this happens, try to help students channel their ideas toward a more realistic and achievable goal. As you move around the classroom, notice what students are talking about and try to help them maintain the connection between science fact and science fiction. Be available for consultation.

1. Make sure your students fully explore the ways in which they find out about science information on a daily basis, both formal and informal. Help them distinguish between different levels of information and the vehicles used to communicate it.
2. Ensure the list is as comprehensive as possible and that it can be added to at any time during the investigation. Help students distinguish between science fiction and science fact within the list of items.
3. Make sure that all students fully understand the task ahead of them and also how this will relate to their assessment.
4. It is probably best to have a range of science fiction items available for students to study. You need to be practical here. Showing a 120-minute movie in class is probably not the best use of time. But, you could encourage students to do this in the evenings or at the weekend. You and your students could assemble a number of clips from a variety of science fiction movies. This would also be useful for your future use of this module. Another idea might be to set up a display table in the classroom at which a collection of science fiction examples is displayed for the duration of the investigation, which you can add to in the future.
5. Encourage students to be creative in their plans. Emphasize that the communication piece is for a general public audience and must have a strong basis in astronomy.

Investigation 8: Discovering the Difference Between Science Fact and Science Fiction

a) Why do you think the piece blends fiction with fact?

b) What do the producers of the piece seem to want the readers/viewers to believe?

c) How could you change the piece to make it more factual?

6. Use all of the resources in your classroom, plus your journals, to decide what content will be in your piece.

   Divide up the work so that everyone in your group has a fair share of the effort. Some people might be better at artwork, while others are good writers or researchers.

   a) Outline what you want in your piece, and then do the research you need to make sure that the science content is correct. Be as creative as you can, but don't try to cover a huge topic. Some ideas might be:

   - Water on Mars
   - Remote space travel to distant planets
   - Search for extraterrestrial intelligence
   - U.S. space program compared to other countries' programs
   - Space Shuttle flights
   - New planets
   - The Apollo missions
   - New technologies used for space observation
   - History of astronomy (famous astronomers)
   - Potential space hazards

7. Take your accurate information and decide how you will weave fiction into your piece. It might be useful to limit the fiction to just three or four items, so that your audience isn't overwhelmed.

   a) Create your piece and your presentation. Check with your teacher to make sure that your "accurate" information really is accurate.

8. When everything is ready, your teacher will set up a presentation schedule. You will be responsible both for presenting your piece and analyzing other groups' pieces for fact versus fiction.

Teacher Commentary

6. It may be helpful to have students submit their plans to you for advice and approval. This way, students are less likely to take on more than they can manage, and you will have a better idea of what they are likely to need in terms of resources and advice.
7. You will need to think carefully on how to set up the presentations, especially if there is some variety of media involved. From following students' progress through this investigation, you will have a good sense of what might be needed and be able to plan accordingly.

Assessment Tools

Key Question-Evaluation Sheet
Use this evaluation to help students understand and internalize basic expectations for the warm-up activity. The **Key Question-Evaluation Sheet** emphasizes that you want to see evidence of prior knowledge and that students should communicate their thinking clearly. You will not likely have time to apply this assessment every time students complete a warm-up activity; yet, in order to ensure that students value committing their initial conceptions to paper and are taking the warm up seriously, you should always remind them of the criteria. When time permits, use this evaluation sheet as a spot check on the quality of their work.

Journal Entry Evaluation-Sheet
Use this sheet as a general guideline for assessing student journals, adapting it to your classroom if desired. You should give the **Journal Entry-Evaluation Sheet** to students early in the module, discuss it with them, and use it to provide clear and prompt feedback.

Journal Entry-Checklist
Use this checklist as a guide for quickly checking the quality and completeness of journal entries.

INVESTIGATING EARTH IN SPACE: ASTRONOMY

Review and Reflect

Review

1. How did other groups capture your interest with science fact?
2. How did they capture your interest with science fiction?

Reflect

3. What use do you think science fiction has in the world?
4. How can you tell science fact from fiction in your everyday life? What would you look for? What resources could you use to help you figure out fact from fiction?

Thinking about the Earth System

5. In what ways can the factual parts of the Earth System combine with fantasy to make science fiction?

Thinking about Scientific Inquiry

6. In which parts of the investigation did you:
   a) Analyze information?
   b) Compare ideas?
   c) Revise your ideas?
   d) Use your imagination?
   e) Share ideas with others?
   f) Organize information?
   g) Pull your information together to make a presentation?

Evidence for Ideas

Teacher Commentary

Review and Reflect

Review
Give your students ample time to review what they have learned. Help them see that while an investigation like this provides some answers to scientific question, they often raise further questions. Have them consider whether science fiction is helpful or harmful in this respect, and if it may be a way for the general public to learn real science or not.

1. Have students discuss the items produced by other groups in a carefully considered way.
2. Here students can respond to the level and relative success of the five ideas involved.

Reflect
3. Allow students to consider this question carefully. Have them back up their ideas and opinions with examples and evidence if possible.
4. It is important that your students develop their critical thinking skills, not just for their science studies, but for life management. There is an opportunity here to help them see the difference between science fiction and science misinformation. Have them focus on the work they have done and how they were careful to use reliable sources of information. Emphasize the role of credible evidence in science.

Thinking about the Earth System
5. Science fiction depends upon science fact to be convincing. It can take the ideas we have about the Earth System and extend or distort them to make a compelling adventure science fiction story.

Thinking about Scientific Inquiry
To help students understand the relevance of these processes to their lives, ask them to think of everyday examples of when they use these processes.

6. Answers will vary.

> **Assessing the Final Investigation**
>
> Students' work throughout the module culminates with this final investigation. To complete it, students need a working knowledge of previous activities. The last investigation is good review and chance to demonstrate proficiency because it refers to the previous steps. For an idea on using the last investigation as a performance-based exam, see the section in the back of this Teacher's Edition. If you chose to use a scoring guide, review it with students before they begin their work.
>
> **Assessment Opportunity**
>
> Comparisons between students' initial answers to these questions (in the pre-assessment at the beginning of the module) and those they are now able to give provide valuable data for assessment.

Teacher Review

Use this section to reflect on and review the investigation. Keep in mind that your notes here are likely to be especially helpful when you teach this investigation again. Questions listed here are examples only.

Student Achievement

What evidence do you have that all students have met the science content objectives?

_____

Are there any students who need more help in reaching these objectives? If so, how can you provide this?_____

_____

What evidence do you have that all students have demonstrated their understanding of the inquiry processes?_____

_____

Which of these inquiry objectives do your students need to improve upon in future investigations? _____

_____

What evidence do the journal entries contain about what your students learned from this investigation? _____

_____

Planning

How well did this investigation fit into your class time?_____

_____

What changes can you make to improve your planning next time? _____

_____

Guiding and Facilitating Learning

How well did you focus and support inquiry while interacting with students?

_____

What changes can you make to improve classroom management for the next investigation or the next time you teach this investigation? _____

_____

Teacher Review

How successful were you in encouraging all students to participate fully in science learning? _____

_____

How did you encourage and model the skills values, and attitudes of scientific inquiry? _____

_____

How did you nurture collaboration among students? _____

_____

Materials and Resources

What challenges did you encounter obtaining or using materials and/or resources needed for the activity? _____

_____

What changes can you make to better obtain and better manage materials and resources next time? _____

_____

Student Evaluation

Describe how you evaluated student progress. What worked well? What needs to be improved? _____

_____

How will you adapt your evaluation methods for next time? _____

_____

Describe how you guided students in self-assessment. _____

_____

Self Evaluation

How would you rate your teaching of this investigation? _____

_____

What advice would you give to a colleague who is planning to teach this investigation? _____

_____

Reflecting

Back to the Beginning
How have your ideas about the Earth in Space changed from when you started? Look at following items carefully. Draw and write down the ideas you have now about these items. How have your ideas changed?

- Your original sketch of the Solar System with labels.
- Your explanation of what gravity is.
- Your list of all the objects you know about that are outside the Solar System.
- The explanation of two of the items from your list of objects in space and the explanation of what they are.

Thinking about the Earth System
The investigations in this module have had you looking at the Earth, the Solar System, and beyond. Think about the idea of "systems within systems." Answer the following question in your journal:

- What connections can you make between Earth in Space and the Earth System?

Thinking about Scientific Inquiry
Review the investigations you have done and the inquiry processes you have used. Answer the following questions in your journal:

- What scientific processes did you use?
- How did scientific inquiry processes help you learn about Earth in Space?

A New Beginning
This investigation into Earth in Space and astronomy is now complete, but that's not the end of the story! As time goes by, you will see, and hear about, many new space-science events and discoveries. Maybe you will actually travel in space one day. Be alert for opportunities to add to your knowledge and understanding.

Reflecting

This is the point at which your students review what they have learned throughout the module. This review is very important. Allow students time to work on this in a thoughtful way.

Back to the Beginning

Encourage students to complete this final review without looking at their journal entries from the beginning of the module. Their initial entries may influence their responses.

When students have completed their writing, encourage them to revisit their initial answers from the pre-assessment. Compare their writings at the end of the unit to their writings at the beginning. It is important that students not be left with the impression that they now know all there is to know about Earth in space and astronomy. Emphasize that learning is a continuous process throughout their lives, and that practicing scientists themselves are still faced with a host of uncertainties and unanswered questions about our Solar System.

> **Assessment Opportunity**
> Comparisons between students' initial answers to these questions (in the pre-assessment at the beginning of the module) and those they are now able to give provide valuable data for assessment.

Thinking about the Earth System

Now that your students are at the end of this module, ask them to make connections between the Earth System and Earth in space. You may want to do this through a concept map. This is an opportunity for you to gauge how well students have developed their understanding of the Earth System for assessment purposes.

Thinking about Scientific Inquiry

To help students understand the relevance of these processes to their lives, ask them to think of everyday examples of when they use these processes (finding out where a misplaced book has gone; forming an opinion about a new TV show; winning an argument).

NOTES

Appendices

Investigating Earth in Space: Astronomy Alternative End-of-Module Assessment

Part A. Matching.
Write the letter of the term from column B that matches the description in Column A.

Column A	Column B
1. Solar System	A. 28.5 day period during which the Moon orbits the Earth
2. galaxy	B. Process on the Sun whereby hydrogen nuclei fuse to form helium nuclei
3. gravity	C. Huge clouds of gas and dust that give rise to stars
4. lunar cycle	D. Bodies of metallic and rocky material that orbit the Sun
5. nebula	E. Body of ice, rock or dust, loosely packed together, that orbits the Sun
6. asteroid	F. Collection of celestial bodies orbiting our Sun.
7. comet	G. Large group of stars, dust and gas held together by gravity
8. Astronomical Unit	H. Force that attracts all matter to all other matter in the universe
9. light-year	I. 150 million kilometers
10. solar fusion	J. Distance light travels in one year

Part B: Multiple Choice.
Provide the letter of the choice that best answers the questions or completes the statement.

11. Galaxies are classified as:
    A. spiral
    B. lenticular
    C. elliptical
    D. irregular
    E. all of the above

12. Our galaxy, the Milky Way, is what type of galaxy?
    A. spiral
    B. lenticular
    C. elliptical
    D. irregular
    E. nebular

13. Mars is which planet, counting out from the Sun?
    A. first
    B. second
    C. third
    D. fourth
    E. fifth
14. The process on our Sun that produces energy is:
    A. cold fusion
    B. nuclear fission
    C. nuclear fusion
    D. photosynthesis
    E. phototropism

15. The pull of gravity on the Moon is:
    A. The same as on the Earth
    B. Six times greater than on the Earth
    C. Twice that of the Earth
    D. One-sixth that of the Earth
    E. None of the above

16. Astronomers who study the origin and explanation of the universe are called:
    A. chemists
    B. constellationists
    C. cosmonauts
    D. cosmologists
    E. astronauts

17. The life cycle of a star is completely dependent upon the star's:
    A. location
    B. mass
    C. color
    D. composition
    E. galaxy

18. The outermost part of the Sun is called the:
    A. chromosphere
    B. photosphere
    C. corona
    D. core
    E. mantle

19. Which of the planets below has the shortest year (in Earth days)?
   A. Pluto
   B. Earth
   C. Jupiter
   D. Venus
   E. Mercury

20. The inner and outer cores of the Earth are made mostly of:
   A. Iron
   B. Lead
   C. Gold
   D. Titanium
   E. Mercury

Assessment Answers

Answers

Part A
1. f
2. g
3. h
4. a
5. c
6. d
7. e
8. i
9. j
10. b

Part B
11. e
12. a
13. d
14. c
15. d
16. d
17. b
18. c
19. e
20. a

Investigating Earth Systems Assessment Tools

Assessing the Student *IES* Journal

- Journal Entry-Evaluation Sheet
- Journal Entry-Checklist
- Key Question-Evaluation Sheet
- Investigation Journal Entry-Evaluation Sheet
- Review and Reflect Journal Entry-Evaluation Sheet

Assisting Students with Self Evaluation

- Group Participation Evaluation Sheet I
- Group Participation Evaluation Sheet II

Assessing the Final Investigation

- Final Investigation Evaluation Sheet
- Student Presentation Evaluation Form

References

- Doran, R., Chan, F., and Tamir, P. (1998). *Science Educator's Guide to Assessment.*
- Leonard, W.H., and Penick, J.E. (1998). *Biology – A Community Context.* South-Western Educational Publishing. Cincinnati, Ohio.

Journal Entry-Evaluation Sheet

Name: _____ Date: _____ Module: _____

Explanation: The journal is an important component of each *IES* module. In using the journal as you investigate Earth science questions, you are mirroring what scientists do. The criteria, along with others that your teacher may add, will be used to evaluate the quality of your journal entries. Use these criteria, along with instructions within investigations, as a guide.

Criteria

1. Entry Made
      1   2   3   4   5   6   7   8   9   10   _____
   Blank   Nominal   Above average Thorough

2. Detail
      1   2   3   4   5   6   7   8   9   10   _____
   Few dates   Half the time   Most days   Daily
   Little detail   Some detail   Good detail   Excellent detail

3. Clarity
      1   2   3   4   5   6   7   8   9   10   _____
   Vague   Becoming clearer   Clearly expressed
   Disorganized   well organized

4. Data Collection/Analysis
      1   2   3   4   5   6   7   8   9   10   _____
   Data collected   Data collected,   Data collected
   Not analyzed   some analyzed   and analyzed

5. Originality
      1   2   3   4   5   6   7   8   9   10   _____
   Little evidence   Some evidence   Strong evidence
   of originality   of originality   of originality

6. Reasoning/Higher-Order Thinking
      1   2   3   4   5   6   7   8   9   10   _____
   Little evidence   Some evidence   Strong evidence
   of thoughtfulness of thoughtfulness of thoughtfulness

7. Other
      1   2   3   4   5   6   7   8   9   10   _____

8. Other
      1   2   3   4   5   6   7   8   9   10   _____

Journal Entry-Checklist

Name: _____ Date: _____ Module: _____

Explanation: The journal is an important component of each *IES* module. In using the journal as you investigate Earth science questions, you are mirroring what scientists do. The criteria, along with others that your teacher may add, will be used to evaluate the quality of your journal entries. Use these criteria, along with instructions within investigations, as a guide.

Criteria

1. Makes entries _____

2. Provides dates and details _____

3. Entry is clear and organized _____

4. Shows data collected _____

5. Analyzes data collected _____

6. Shows originality in presentation _____

7. Shows evidence of higher-order thinking _____

8. Other _____

9. Other _____

Total Earned _____

Total Possible _____

Comments:

Key Question-Evaluation Sheet

Name: _____ Date: _____ Module: _____

	No Entry		Fair		Strong
Shows evidence of prior knowledge	0	1	2	3	4
Reflects discussion with classmates	0	1	2	3	4

Additional Comments

Key Question-Evaluation Sheet

Name: _____ Date: _____ Module: _____

	No Entry		Fair		Strong
Shows evidence of prior knowledge	0	1	2	3	4
Reflects discussion with classmates	0	1	2	3	4

Additional Comments

Key Question-Evaluation Sheet

Name: _____ Date: _____ Module: _____

	No Entry		Fair		Strong
Shows evidence of prior knowledge	0	1	2	3	4
Reflects discussion with classmates	0	1	2	3	4

Additional Comments

Investigation Journal Entry-Evaluation Sheet

Name: _____ Date: _____ Module: _____

Criteria

1. Completeness of written investigation
      1 2 3 4 5 6 7 8 9 10   _____
   Blank   Incomplete   Thorough

2. Participation in investigations
      1 2 3 4 5 6 7 8 9 10   _____
   None or little;   Needs minimal guidance, Leads, is inquisitive,
   unable to guide   sometimes helping others persistent, focused
   self

3. Skills attained
      1 2 3 4 5 6 7 8 9 10   _____
   Few skills   Tends to use some   High degree of
   evident   appropriate skills   appropriate skills used

4. Investigation Design
      1 2 3 4 5 6 7 8 9 10   _____
   Variables not   Sometimes   Considers variables
   considered   considers variables,   Sound rationale for
   techniques   uses logical techniques techniques
   illogical

5. Conceptual understanding of content
      1 2 3 4 5 6 7 8 9 10   _____
   No evidence   Approaches understanding Exceeds expectations
   of understanding of most concepts   for content attainment

6. Ability to explain/discuss inquiry
      1 2 3 4 5 6 7 8 9 10   _____
   Unable to   Some ability to   Uses scientific reasoning
   articulate   explain/discuss   to explain any
   scientific thought the inquiry   aspect of the inquiry

7. Other
      1 2 3 4 5 6 7 8 9 10   _____

8. Other
      1 2 3 4 5 6 7 8 9 10   _____

Review and Reflect Journal Entry-Evaluation Sheet

Name: _____ Date: _____ Module: _____

Criteria	Blank		Fair			Excellent
Thoroughness of answers	0	1	2	3	4	5
Content of answers	0	1	2	3	4	5
Other	0	1	2	3	4	5

Review and Reflect Journal Entry-Evaluation Sheet

Name: _____ Date: _____ Module: _____

Criteria	Blank		Fair			Excellent
Thoroughness of answers	0	1	2	3	4	5
Content of answers	0	1	2	3	4	5
Other	0	1	2	3	4	5

Review and Reflect Journal Entry-Evaluation Sheet

Name: _____ Date: _____ Module: _____

Criteria	Blank		Fair			Excellent
Thoroughness of answers	0	1	2	3	4	5
Content of answers	0	1	2	3	4	5
Other	0	1	2	3	4	5

Group Participation Evaluation Sheet I

Key:
4 = Worked on his/her part and assisted others
3 = Worked on his/her part
2 = Worked on part less than half the time
1 = Interfered with the work of others
0 = No work

My name is _____ .   I give myself a _____

The other people in my group are: I give each person:

A. _____   _____

B. _____   _____

C. _____   _____

D. _____   _____

Key:
4 = Worked on his/her part and assisted others
3 = Worked on his/her part
2 = Worked on part less than half the time
1 = Interfered with the work of others
0 = No work

My name is _____ .

The other people in my group are:

A. _____

B. _____

C. _____

D. _____

Group Participation Evaluation Sheet II

Name: _____ Date: _____ Module: _____

Key:
Highest rating _____
Lowest rating _____

1. In the chart, rate each person in your group, including yourself.

	Names of Group Members				
Quality of Work					
Quantity of Work					
Cooperativeness					
Other Comments					

2. What went well in your investigation?

3. If you could repeat the investigation, how would you change it?

Investigating Earth in Space: Astronomy

Final Investigation Evaluation Sheet

Alerting students

Before your students begin the final investigation, they must understand what is expected of them and how they will be evaluated on their performance. Review the task thoroughly, setting time guidelines and parameters (whom they may work with, what materials they can use, etc.). Spell out the evaluation criteria for each level of proficiency shown below. Use three categories for a 3-point scale (Achieved, Approaching, Attempting). If you prefer a 5-point scale, add the final two categories.

Name: _____ Date: _____ Module: _____

	Understanding of concepts and inquiry	Use of evidence to explain and support results	Communication of ideas	Thoroughness of work
Exceeding proficiency 5	Demonstrates complete and unambiguous understanding of the problem and inquiry processes used.	Uses all evidence from inquiry that is factually relevant, accurate, and consistent with explanations offered.	Communicates ideas clearly and in a compelling and elegant manner to the intended audience.	Goes beyond all deliverables agreed upon for the project and has extended the data collection and analysis.
Achieved proficiency 4	Demonstrates fairly complete and reasonably clear understanding of the problem and inquiry processes used.	Uses the major evidence from inquiry that is relevant and consistent with explanations offered.	Communicates ideas clearly and coherently to the intended audience.	Includes all of the deliverables agreed upon for the project.
Approaching proficiency 3	Demonstrates general, yet somewhat limited understanding of the problem and inquiry processes used.	Uses evidence from inquiry to support explanations but may mix fact with opinion, omit significant evidence, or use evidence that is not totally accurate.	Completes the task satisfactorily but communication of ideas is incomplete, muddled, or unclear.	Work largely complete but missing one of the deliverables agreed upon for the project.
Attempting proficiency 2	Demonstrates only a very general understanding of the problem and inquiry processes used.	Uses generalities or opinion more than evidence from inquiry to support explanations.	Communication of ideas is difficult to understand or unclear.	Work missing several of the deliverables agreed upon for the project.
Non-proficient 1	Demonstrates vague or little understanding of the problem and inquiry processes used.	Uses limited evidence to support explanations or does not attempt to support explanations.	Communication of ideas is brief, vague, and/or not understandable.	Work largely incomplete; missing many of the deliverables agreed upon for the project.

Student Presentation Evaluation Form

Student Name_____ Date_____

Topic_____

	Excellent	Fair		Poor
Quality of ideas	4	3	2	1
Ability to answer questions	4	3	2	1
Overall comprehension	4	3	2	1

COMMENTS_____
_____
_____
_____
_____

Student Presentation Evaluation Form

Student Name_____ Date_____

Topic_____

	Excellent	Fair		Poor
Quality of ideas	4	3	2	1
Ability to answer questions	4	3	2	1
Overall comprehension	4	3	2	1

COMMENTS_____
_____
_____
_____
_____

Blackline Master *Earth in Space: Astronomy* P. 1

Astronomy Items

- **Draw a sketch of the Solar System and label as many objects as you can in it.**

- **Explain what gravity is.**

- **Make a list of all the objects you know about that are outside the Solar System.**

- **Choose two items from your list of objects in space and explain what they are.**

Use with *Earth in Space: Astronomy* Pre-assessment.

Blackline Master *Earth in Space: Astronomy* **P. 2**

Student Journal Cover Sheet
Investigating Earth in Space: Astronomy

Name: _____

Group Members:

1. _____

2. _____

3. _____

4. _____

Teacher: _____

Class: _____

Dates of Investigation:

Start _____ Complete _____

Keep this journal with you at all times during your study of
Investigating Earth in Space: Astronomy.

Use with *Earth in Space: Astronomy* Pre-assessment.

Blackline Master *Earth in Space*: Astronomy I. 1

Name: _____

Earth System Connection Sheet

When you finish an investigation, use this sheet to record any links you can make with the Earth system. By the end of the module you should have as complete a diagram as possible.

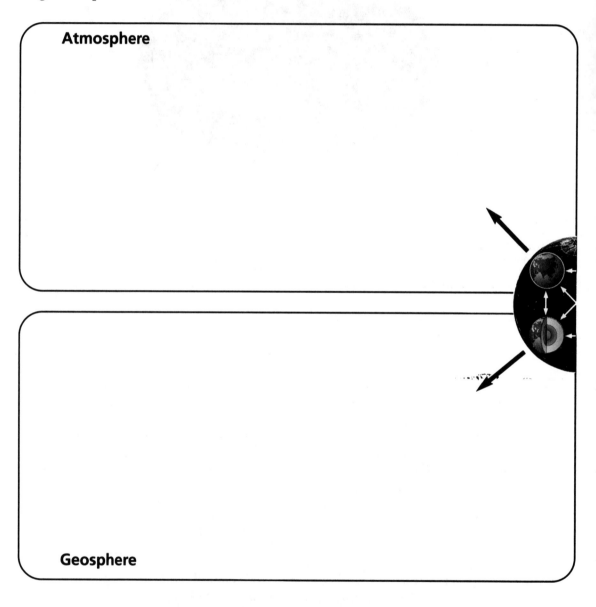

Atmosphere

Geosphere

Use with *Earth in Space: Astronomy* Introducing the Earth System.

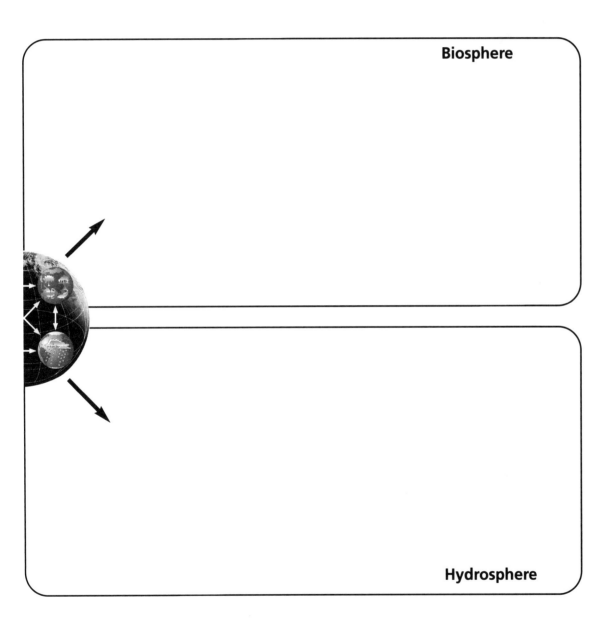

Blackline Master *Earth in Space: Astronomy* **1.2**

Inquiry Processes

 • Explore questions to answer by inquiry.

 • Design an investigation.

 • Conduct an investigation.

 • Collect and review data using tools.

 • Use evidence to develop ideas.

 • Consider evidence for explanations.

 • Seek alternative explanations.

 • Show evidence and reasons to others.

 • Use mathematics for science inquiry.

Use with *Earth in Space: Astronomy* Introducing the Earth System.

Blackline Master *Earth in Space: Astronomy* **3.1**

Data Table - Mini-Investigation B			
Jumper's Name	Earth Jump Distance (cm)	Moon Jump Distance (cm)	Jupiter Jump Distance (cm)

Use with *Earth in Space: Astronomy* Investigation 3: The Earth and Its Moon

NOTES

NOTES

NOTES

NOTES

NOTES